S0-BNC-276

DATE DUE

Unless Recalled Earlier

DEMCO 38-297

From Cologne to Chapel Hill

PROFILES, PATHWAYS, AND DREAMS
Autobiographies of Eminent Chemists

Jeffrey I. Seeman, Series Editor

From Cologne to Chapel Hill

Ernest L. Eliel

RECEIVED

DEC - 1 1993

MSU - LIBRARY

American Chemical Society, Washington, DC 1990

QD
22
.E53
A3
1990

Library of Congress Cataloging-in-Publication Data

Eliel, Ernest Ludwig, 1921–
  From Cologne to Chapel Hill.

  (Profiles, pathways, and dreams)
  Includes bibliographical references.

  1. Eliel, Ernest Ludwig, 1921–   . 2. Chemists—
United States—Biography. 3. Chemistry, Organic—
United States—History—20th century.

  I. Title. II. Series.

QD22.E53A3 1990 540'.92 [B]   90–255
ISBN 0–8412–1767–X (cloth)
ISBN 0–8412–1793–9 (pbk.)   CIP

The paper used in this publication meets the minimum requirements of American
National Standard for Information Sciences—Permanence of Paper for Printed Library
Materials, ANSI Z39.48–1984.
∞

Copyright © 1990

American Chemical Society

All Rights Reserved. The appearance of the code at the bottom of the first page of each
chapter in this volume indicates the copyright owner's consent that reprographic copies
of the chapter may be made for personal or internal use or for the personal or internal
use of specific clients. This consent is given on the condition, however, that the copier
pay the stated per-copy fee through the Copyright Clearance Center, Inc., 27 Congress
Street, Salem, MA 01970, for copying beyond that permitted by Sections 107 or 108 of
the U.S. Copyright Law. This consent does not extend to copying or transmission by
any means—graphic or electronic—for any other purpose, such as for general
distribution, for advertising or promotional purposes, for creating a new collective
work, for resale, or for information storage and retrieval systems. The copying fee for
each chapter is 1047–8329/90/$00.00+.75.

The citation of trade names and/or names of manufacturers in this publication is not to
be construed as an endorsement or as approval by ACS of the commercial products or
services referenced herein; nor should the mere reference herein to any drawing,
specification, chemical process, or other data be regarded as a license or as a
conveyance of any right or permission to the holder, reader, or any other person or
corporation, to manufacture, reproduce, use, or sell any patented invention or
copyrighted work that may in any way be related thereto. Registered names,
trademarks, etc., used in this publication, even without specific indication thereof, are
not to be considered unprotected by law.

PRINTED IN THE UNITED STATES OF AMERICA

Profiles, Pathways, and Dreams

Jeffrey I. Seeman, *Series Editor*

M. Joan Comstock, *Head, ACS Books Department*

1990 ACS Books Advisory Board

Paul S. Anderson
Merck Sharp & Dohme
  Research Laboratories

V. Dean Adams
Tennessee Technological
  University

Alexis T. Bell
University of California—
  Berkeley

Malcolm H. Chisholm
Indiana University

Natalie Foster
Lehigh University

G. Wayne Ivie
U.S. Department
  of Agriculture, Agricultural
  Research Service

Mary A. Kaiser
E. I. du Pont de Nemours
  and Company

Michael R. Ladisch
Purdue University

John L. Massingill
Dow Chemical Company

Robert McGorrin
Kraft General Foods

Daniel M. Quinn
University of Iowa

Elsa Reichmanis
AT&T Bell Laboratories

C. M. Roland
U.S. Naval Research Laboratory

Stephen A. Szabo
Conoco Inc.

Wendy A. Warr
Imperial Chemical Industries

Robert A. Weiss
University of Connecticut

Foreword

In 1986, the ACS Books Department accepted for publication a collection of autobiographies of organic chemists, to be published in a single volume. However, the authors were much more prolific than the project's editor, Jeffrey I. Seeman, had anticipated, and under his guidance and encouragement, the project took on a life of its own. The original volume evolved into 22 volumes, and the first volume of *Profiles, Pathways, and Dreams: Autobiographies of Eminent Chemists* was published in 1990. Unlike the original volume, the series was structured to include chemical scientists in all specialties, not just organic chemistry. Our hope is that those who know the authors will be confirmed in their admiration for them, and that those who do not know them will find these eminent scientists a source of inspiration and encouragement, not only in any scientific endeavors, but also in life.

M. Joan Comstock
Head, Books Department
American Chemical Society

Contributors

We thank the following corporations and Herchel Smith for their generous financial support of the series Profiles, Pathways, and Dreams.

Akzo nv

Bachem Inc.

E. I. du Pont de Nemours and Company

Duphar B.V.

Eisai Co., Ltd.

Fujisawa Pharmaceutical Co., Ltd.

Hoechst Celanese Corporation

Imperial Chemical Industries PLC

Kao Corporation

Mitsui Petrochemical Industries, Ltd.

The NutraSweet Company

Organon International B.V.

Pergamon Press PLC

Pfizer Inc.

Philip Morris

Quest International

Sandoz Pharmaceuticals Corporation

Sankyo Company, Ltd.

Schering–Plough Corporation

Shionogi Research Laboratories, Shionogi & Co., Ltd.

Herchel Smith

Suntory Institute for Bioorganic Research

Takeda Chemical Industries, Ltd.

Takasago International Corporation

Unilever Research U.S., Inc.

About the Editor

JEFFREY I. SEEMAN received his B.S. with high honors in 1967 from the Stevens Institute of Technology in Hoboken, New Jersey, and his Ph.D. in organic chemistry in 1971 from the University of California, Berkeley. Following a two-year staff fellowship at the Laboratory of Chemical Physics of the National Institutes of Health in Bethesda, Maryland, he joined the Philip Morris Research Center in Richmond, Virginia, where he is currently a senior scientist and project leader. In 1983–1984, he enjoyed a sabbatical year at the Dyson Perrins Laboratory in Oxford, England, and claims to have visited more than 90% of the castles in England, Wales, and Scotland.

Seeman's 80 published papers include research in the areas of photochemistry, nicotine and tobacco alkaloid chemistry and synthesis, conformational analysis, pyrolysis chemistry, organotransition metal chemistry, the use of cyclodextrins for chiral recognition, and structure–activity relationships in olfaction. He was a plenary lecturer at the Eighth IUPAC Conference on Physical Organic Chemistry held in Tokyo in 1986 and has been an invited lecturer at numerous scientific meetings and universities. Currently, Seeman serves on the Petroleum Research Fund Advisory Board. He continues to count Nero Wolfe and Archie Goodwin among his best friends.

Contents

Photographs

Preface

"HOW DID YOU GET THE IDEA—and the good fortune—to convince 22 world-famous chemists to write their autobiographies?" This question has been asked of me, in these or similar words, frequently over the past several years. I hope to explain in this preface how the project came about, how the contributors were chosen, what the editorial ground rules were, what was the editorial context in which these scientists wrote their stories, and the answers to related issues. Furthermore, several authors specifically requested that the project's boundary conditions be known.

As I was preparing an article[1] for *Chemical Reviews* on the Curtin–Hammett principle, I became interested in the people who did the work and the human side of the scientific developments. I am a chemist, but I have a deep appreciation of history, especially in the sense of individual accomplishments. Readers' responses to the historical section of that review encouraged me to take an active interest in the history of chemistry. The concept for Profiles, Pathways, and Dreams resulted from that interest.

My goal for Profiles was to document the development of modern organic chemistry by having individual chemists discuss their roles in this development. Authors were not chosen to represent my choice of the world's "best" organic chemists, as one might choose the "baseball all-star team of the century". Such an attempt would be foolish: Even the selection committees for the Nobel prizes do not make their decisions on such a premise.

The selection criteria were numerous. Each individual had to have made seminal contributions to organic chemistry over a multi-decade career. (The average age of the authors is over 70!) Profiles would represent scientists born and professionally productive in different countries. (Chemistry in 13 countries is detailed.) Taken together, these individuals were to have conducted research in nearly all subspecialties of organic chemistry. Invitations to contribute were based on solicited advice and on recommendations of chemists from five continents, including nearly all of the contributors. The final assemblage was selected entirely and exclusively by me. Not all who were invited chose to participate, and not all who should have been invited could be asked.

A very detailed four-page document was sent to the contributors, in which they were informed that the objectives of the series were

1. to delineate the overall scientific development of organic chemistry during the past 30–40 years, a period during which this field has dramatically changed and matured;

2. to describe the development of specific areas of organic chemistry; to highlight the crucial discoveries and to examine the impact they have had on the continuing development in the field;

3. to focus attention on the research of some of the seminal contributors to organic chemistry; to indicate how their research programs progressed over a 20–40-year period; and

4. to provide a documented source for individuals interested in the hows and whys of the development of modern organic chemistry.

One noted scientist explained his refusal to contribute a volume by saying, in part, that "it is extraordinarily difficult to write in good taste about oneself. Only if one can manage a humorous and light touch does it come off well. Naturally, I would like to place my work in what I consider its true scientific perspective, but . . ."

Each autobiography reflects the author's science, his lifestyle, and the style of his research. Naturally, the volumes are not uniform, although each author attempted to follow the guidelines. "To write in good taste" was not an objective of the series. On the contrary, the authors were specifically requested not to write a review article of their field, but to detail their own research accomplishments. To the extent that this instruction was followed and the result is not "in good taste", then these are criticisms that I, as editor, must bear, not the writer.

As in any project, I have a few regrets. It is truly sad that Egbert Havinga, who wrote one volume, and David Ginsburg, who translated another, died during the development of this project. There have been many rewards, some of which are documented in my personal account of this project, entitled "Extracting the Essence: Adventures of an Editor" published in CHEMTECH.[2]

Acknowledgments

I join the entire chemical community in offering each author unbounded thanks. I thank their families and their secretaries for their contributions. Furthermore, I thank numerous chemists for reading and reviewing the chapters, for lending photographs, for sharing information, and for providing each of the authors and me the encouragement to proceed in a project that was far more costly in time and energy than any of us had anticipated.

I thank my employer, Philip Morris USA, and J. Charles, R. N. Ferguson, K. Houghton, and W. F. Kuhn, for without their support, Profiles, Pathways, and Dreams could not have been. I thank ACS Books, and in particular, Robin Giroux (acquisitions editor), Karen Schools Colson (production manager), Janet Dodd (senior editor), Joan Comstock (department head), and their staff for their hard work, dedication, and support. Each reader no doubt joins me in thanking 23 corporations and Herchel Smith for financial support for the project.

I thank my wife Suzanne, for she assisted Profiles in both practical and emotional ways. I thank my children Jonathan and Brooke for their patient support and understanding; remarkably, I have been working on Profiles for more than half of their lives—probably the only half that they can remember! My family hardly knows a husband or father who doesn't live the life of an editor. Finally, I again thank all those mentioned and especially my family, friends, colleagues, and the 22 authors for allowing me to share this experience with them.

JEFFREY I. SEEMAN
Philip Morris Research Center
Richmond, VA 23234

February 15, 1990

[1] Seeman, J. I. *Chem. Rev.* **1983**, *83*, 83–134.
[2] Seeman, J. I. *CHEMTECH* **1990**, 20(2), 86–90.

Editor's Note

"Dedicated to Professor Ernest L. Eliel, to commemorate his 65th birthday and to honor his commitments to science, education, society, and professionalism." [*J. Am. Chem. Soc.* **1987**, *109*, 3453]

We each have our favorite publications, those of which we are especially proud, for which we feel we have made a contribution using our maximum efforts and capabilities. In 1987, my colleagues and I dedicated one such paper to Eliel, and the quote is repeated here, for without him, the series Profiles, Pathways, and Dreams might not have been.

Eliel is strongly committed to science, education, society, professionalism, and his colleagues. One frequently finds an acknowledgment to Eliel in a paper, sometimes due to scientific input, at other times to less scientific but essential assistance, such as grammatical or translational help for non-English speaking authors. Eliel encouraged my writing a review of the Curtin–Hammett principle and later suggested that I publish it in *Chemical Reviews*. In that article, I wrote a historical account of the development of the chemistry, and Eliel kindly shared his personal story. Those experiences led me, following a chain of events, to the Profiles series. Eliel was one of the first to commit himself to write a volume.

Telling one's story—whether it be a brief account of a single idea published 35 years previously as the Winstein–Holness/Eliel–Ro equation or an autobiography of one's entire scientific and professional career—can frequently surprise the writer. I well remember Eliel's comments after examining my *Chemical Reviews* manuscript; he was quite surprised at how strongly and vehemently some of his emotions rang through his quotes.

Eliel is a clear and precise thinker. His pedagogical instincts and organizational capabilities have manifested themselves in the writing of important textbooks and reviews. The last three decades of organic chemists have grown up on *Stereochemistry of Carbon Compounds* (1962), for which a successor book will soon be published.

It is quite true that most organic chemists associate Eliel with the themes of stereochemistry and conformational analysis. As detailed in his volume, Eliel's research flows naturally, year after year, spurred on by the availability of new instrumentation and new methodologies. Few people think of Eliel as a synthetic organic chemist; he certainly doesn't characterize himself in that way. Yet, one consequence of his stereochemical investigations was the discovery of various stereoselective electrophilic reactions of 2-lithio-1,3-dithianes. Subsequently, a variety of enantioselective syntheses were developed in his laboratories.

How does a man manage to do so much? I recall a telephone conversation with Eliel many years ago. I had called for two reasons: first, to ask him to look for an undergraduate student for a summer position; second, to obtain his opinion on some complicated chemistry. He was very busy, as he usually is, but he did not act or sound rushed. We got to the point quickly but not feverishly so. As I began to say goodbye, he stopped me. "Let me read the announcement to you," he said. And then Eliel read a detailed description of the summer position, composed while we were talking chemistry, now ready for posting on the University of North Carolina's bulletin board: Eliel's efficiency at the maximum!

Eliel has also involved himself heavily in a variety of professional activities, culminating with two terms on the board of directors of the American Chemical Society, including three years as board chairman. It is remarkable that an individual who escaped the Nazis and earned his

Directors William Nevill, George Pimentel, and Ernest L. Eliel wearing National Chemistry Day T-shirts at the 193rd ACS National Meeting, April 1987, Denver, Colorado. (Photo courtesy of C&E News.)

undergraduate degree during the War in Cuba would serve as chairman of the board of the American Chemical Society. Did his experiences as a youth give Eliel the vigor and determination to achieve or was he able to survive because he was born so strong? "One of my colleagues once said that I was a bulldog," he relates. But it is more than that, for Eliel can be quite shy. He has also been characterized as "a gentle ram" because he is quiet but purposeful. One can always count on Ernest Eliel.

There is no apparent limit to Eliel's willingness to listen and to contribute. Regarding Profiles, he willingly and unhesitatingly added paragraphs and sections, discussion and clarification, to his volume. He offered many suggestions as to potential participants and reviewers. He read Prelog's original text (in German) and examined the translations for accuracy and tone. He offered guidance regarding ACS copyright matters. His was the first volume from which an excerpt was taken for publication in the Beckman Center for the History of Chemistry newsletter.

Many of Eliel's friends, former students, and colleagues joined him at a joint 65th birthday celebration in Chapel Hill in March 1987 for him and Bob Parr. It was a wonderful affair, with lectures and social events. One of the after-dinner speakers, a physical chemist colleague of Parr's, kept referring to Eliel as Ernie. That name hardly suits his personality, for Eliel is called Ernest by his colleagues and friends. Yet the warmth of the man perhaps can best be understood by Eliel's seeking out each of his friends, one by one, to have his banquet program autographed. He approached me with his friendly, perhaps even sheepish, grin. I signed, but only as he signed mine.

From Cologne to Chapel Hill

Ernest L. Eliel

Ernst L. Pleil

Home, Education, and Emigration

I was born into a Jewish family in Cologne, Germany, just before the end of 1921. My father was a lawyer. His family had come from Hesse in the middle of the 19th century, at about the time that Jews were first allowed to live in Cologne. My paternal grandfather, after whom I was named, had been a member of the Cologne town council. My mother's family came from the province of Poznan (now part of Poland); my grandfather, Leonard Tietz, had been the founder of a department store that had become very successful in the western part of Germany.

Early Education

My early youth was uneventful. I attended a Catholic elementary school. All elementary schools in Cologne were denominational; there was no separation of church and state in Germany. The Catholic school happened to be closest to where we lived. In school I was diligent, somewhat introverted, and therefore not very popular. When I was nine, my family decided that I had absorbed all that grade school could give me and that I should enter the *Gymnasium*, the German college-preparatory high school, one year early. Fortunately, the dreaded entrance examination had just been abolished, and I was accepted merely on the basis of my grades and, perhaps, I suspect, my family's standing in the community.

High school proved more challenging. Right from the start I had to study history and geography, as well as Latin. I did not enjoy the

Ernest L. Eliel around 1928.

3

study of Latin, but it unquestionably helped me with modern languages. Moreover, the study of Latin required a logic of thought that prepared me for my later study of science. Two years later, we started biology and French, and the year after that (when I was 12), we studied English and serious mathematics. Physics followed soon thereafter, and chemistry studies started when I was 14. The teaching was challenging throughout, and the school's expectations were high.

It was considered a privilege to study in the *Gymnasium*. Only the top students from the elementary school were admitted to the *Gymnasium*, which was considered the high road to the university. The second tier of elementary school children went to the *Oberrealschule*, which was considered preparatory to a business career, although it did offer the possibility to enter university. The remainder stayed in the elementary school for 8 years and then proceeded to the middle school to learn a trade.

In the *Gymnasium*, those who did not produce would not be promoted and might, eventually, be expelled. The system was very elitist, but the instruction was excellent although, I suspect, less supportive of individual initiative and originality than the American system. For example, I was never asked to undertake a library project. I am not even sure the school had a library!

My father was a highly educated man who valued knowledge, and he was a strong role model for me. When I entered the fourth grade of high school, I had to decide between the *Gymnasium* (where the languages would be Latin and Greek, with some limited continua-

Ernest L. Eliel in the mid-1930s.

tion of French and no teaching of English) and the *Realgymnasium* (where French and English were stressed, Latin was downplayed, and Greek was not taught at all). Of course there was no question that I would pursue the modern-language track, and I told my father so. Although he was prepared for my decision, he, nevertheless, remonstrated with me. When I asked him why I should learn Greek, my father told me that it was essential for any educated person to know the Greek alphabet and to know the Greek language roots found in many modern (especially scientific) terms. Thereupon, we agreed that he would teach me the Greek alphabet and Greek word roots on Sundays, and he did so. Later I got back at him by telling him that his education in the *Gymnasium* had been inadequate in that he had not learned calculus and that an educated person should know calculus! He agreed that I should teach him calculus in a series of Sunday lessons!

We were a tightly knit family, and I was close not only to my parents but also to my two brothers who, having been born before World War I, were much older (by 10 and 13 years) than I (I was born after that war). We played many games together. Some games were just entertaining, but others, such as word games, were educational. As I grew up, I not only became very good at such games but I also became very much aware of my talents. This awareness of my abilities pleased my brothers not at all, and almost from the time I entered school, they concentrated on putting me down and pricking my "balloon" when I tried to act smart. Perhaps partly as a result of that experience and partly because of the attitude of my parents, who were strong believers in the German saying, "*Eigenlob stinkt; Anderlob klingt* (self-praise smells; being praised tells)", I have always been reluctant to toot my own horn.

Encounter with the Nazis

The Nazis came into power when I was 11 and in the third year of high school, and for the next dozen years, they cast an increasingly ominous shadow on my life. I should preface a brief description of events by saying that Cologne, which was inhabited largely by an easy-going and staunchly Catholic population, was far from a hotbed of Nazism. When I was 9 or 10, my aunt had asked me whether I had encountered any antisemitism in school. At that time, I did not know what the word meant.

When Hitler became chancellor in 1933, the situation changed very quickly. The family department store was boycotted and picketed and, within weeks, had to be given up to a consortium of banks at a severe financial loss. Within days, one-third of my class—roughly the proportion whose parents had voted for the National Socialists—

shunned me completely. Within less than a year, that proportion (in a class of less than 30) had increased to two-thirds, because many students had found it opportune to join the Hitler Youth.

Of the third that was still talking to me, about half were children of well-to-do families (who considered Hitler an upstart), and about half were staunch Catholics who were unsympathetic to the Nazi cause. Among the children of well-to-do families were the son and two nephews of Günther Riesen, who had become mayor of Cologne when Konrad Adenauer was unceremoniously ousted from that post in March 1933. Sometime in late 1936, by which time the Nazis had become firmly established, the mayor was suddenly removed overnight. It was rumored that for some reason he had been sent to a concentration camp. This rumor was false; in fact, he had been merely fired for incompetence.

The next day, Riesen's son and two nephews ceased talking to me. One day soon thereafter, when I saw one of them in an empty corridor, he explained to me that it was no longer advisable for him and his group to be seen in my presence. Nonetheless, two of the Catholic pupils continued their friendship to the end; one of them was killed during World War II, and the other is a friend to this day. I must say, however, that the teachers, although some of them were Nazis, never treated me badly, with one exception. Evidently, they respected the fact that I was a good student, especially in mathematics and chemistry.

When the Nazis first came to power, my father, who had fought on the German side as an officer during World War I, considered them a passing phenomenon. That false assumption had vanished by 1935, when the Nuremberg laws, which totally eliminated Jews from Germany's social and cultural life, were passed. That same year, Germany rearmed with impunity, and any external intervention to stop Hitler became increasingly unlikely.

My family decided that I should study in the United States eventually but that I should first complete high school in Germany. The target date for my completion of high school was 1939. Although both my older brothers had left Germany by 1937, my parents felt that they should stay and simply visit us abroad. One reason was that my father, as a lawyer, would find it difficult to find work elsewhere. Another reason was that by 1937 it was impossible to transfer abroad money beyond the sum of about $2000.

Departure from Germany

In late 1937, I registered for a U.S. immigration visa at the U.S. Consulate in Stuttgart. It was too late; I was told that the waiting period would be about 3 years. There were about 600,000 Jews in Germany, many of whom wanted to emigrate to the United States. The U.S. immigration quota for Germans was 27,000 per year, and despite the emergency that existed, no provisions were ever made for the admission of additional refugees, not even after the anti-Jewish pogroms in Germany in November 1938.

By early 1938 another blow fell. Our passports were taken away and were not to be restored unless and until we were cleared for emigration. My father finally decided that we all had to leave as soon as possible. My parents managed to get a visa to go to Palestine, which was then a British mandate. I preferred to continue my schooling in the United Kingdom. In August 1938 I left Germany for good and moved to Edinburgh.

The culture shock of moving to Scotland was not severe. My knowledge of English was quite good because of the excellent training I had had in school, which was supplemented by private lessons and reinforced by a stay in a Welsh summer school the previous year. Moreover, I tended to be quite self-sufficient, even though I had never been separated from my family for any length of time before.

The Scots, although quiet, are kind and warm-hearted people. These traits were particularly true of my landlord and landlady, Ernest and Betty Pennycock, who treated me as a member of the family rather than as a boarder. During my studies at Skerry's College, a school specializing in preparing students for the university entrance examination, I soon found out that I had already learned all that was expected of me, except for the English terminology. (Later, in the university mathematics course, I discovered that this was not true. I was expected to know all about determinants, but I did not.) In March 1939 I passed the Scottish Universities' entrance examination, and in May of that year, I ranked third in a competition for bursaries (scholarships) and was awarded a welcome stipend for the 4 years of university I expected to have ahead of me.

In late spring of 1939, my parents visited me and we toured Scotland. Unfortunately, my father fell seriously ill during this trip; he died 5 months later of endocarditis, which was incurable then. As I spent the summer with my brother in Holland, the storm clouds of war began to thicken. I returned to Edinburgh to study chemistry just days before the outbreak of World War II.

Early Love for Chemistry

I had been in love with chemistry since the age of 11, when my parents had given me a chemistry set and I had found another chemistry set belonging to my brothers. I was further stimulated by a cousin who had a basement chemistry laboratory (but who knew less about chemistry than I did) and by a monthly magazine called *Kosmos*, which described chemical experiments that one could carry out at home, such as dissolving a silver coin in nitric acid and then plating out the silver by reduction. At that time, anyone could buy nitric acid or even concentrated sulfuric acid in a drugstore in Germany; safety considerations were not uppermost in our minds.

Eventually, I began devising my own experiments, and my father became concerned and asked a chemist friend of his for help. It was decreed that I should immediately cease experimenting and, instead, read about chemistry. Thereupon I studied A. F. Holleman's textbooks on inorganic chemistry and organic chemistry, which I greatly enjoyed. An excellent chemistry course in high school (which was taught by a man who was at the same time a Ph.D. candidate at the University of Bonn) complete with demonstrations and laboratory experimentation further deepened my interest and also relieved me of the desire to continue the dangerous experiments at home. By the age of 15, I was set on becoming a chemist.

Life as a Refugee

In Scotland. In 1939 although I was a refugee in Edinburgh, I was technically an enemy alien because I was a German citizen. A British tribunal examined me and certified that I was harmless and to be left alone. Yet on May 12, 1940, as I had nearly finished my first year at the University of Edinburgh—I had taken mathematics, physics, and chemistry—I was interned. After the invasion of Holland and Belgium on May 10, 1940, the British military had become concerned about the many German refugees in Britain and had decided to intern those who, like myself, lived in sensitive areas. Edinburgh was very near the port of Rosyth, the naval base for the British North Sea fleet.

I spent a month in an internment camp near Liverpool and another month in a camp on the Isle of Man. The war had brought my studies to an abrupt halt. In July 1940, after France had fallen and everyone was very nervous about Britain's ability to hold out against the Nazis, we were put on a small boat that took us to the port of Glasgow, where we moored alongside an ocean liner flying the Polish flag. Obviously, the ship had gotten away when the Germans occupied Poland and had been transformed into a troop transport. A bridge was put across our small boat and the liner, and we were invited (but not ordered) to board the big ship. Naturally we asked where it was going, only to be told that its destination was a military secret. I resolved to get away as far as I could and boarded the big ship, which departed shortly thereafter for an unknown destination. Thus I left Europe for good.

In Canada. The ship went to Canada. We were unloaded at Three Rivers, Quebec, and put into a camp that already housed German sailors captured from commercial vessels at the beginning of the war. The German sailors received us by singing antisemitic songs. The situation was clearly untenable, and within a week, we were shipped off to an unfinished camp in New Brunswick; only one water faucet was available for 700 inhabitants. We learned that the British military had told the Canadians that they were sending over "several boatloads of dangerous internees". It took us months to convince the Canadian authorities that we were not dangerous.[1]

Ernest L. Eliel in Canada in 1941.

Life in the Canadian camp, which was eventually built up properly and where I stayed for 10 months, became quite bearable after a while. We spent part of our days cutting trees in what was soon to become a very cold winter, with temperatures down to −35 °F, but we were well clothed and well fed. The camp was full of highly knowledgeable German refugee scientists, who were allowed to teach science courses for the Canadian advanced matriculation examination, which provided entrance into the sophomore year of Canadian universities if one could get out of camp—which was not possible until considerably later. (It also turned out to be impossible to emigrate to the United States from the Canadian camp, even though, as a result of my

1937 application, I had been asked finally to submit papers for immigration in April 1940, just days before I was interned.)

In Cuba. I was eventually released in May 1941 (almost exactly a year after being interned), because my U.S.-based relatives had secured a Cuban visa for me. Through devious ways (via Trinidad and Venezuela) and with considerable trouble en route, I finally arrived in Havana in July 1941. By then, all visa applications had to pass through the State Department in Washington, DC. The fact that I had been interned made my case suspect, and there was a long delay. In December 1941, the United States and Cuba entered World War II, and I was again an enemy alien and therefore an undesirable immigrant. I was stuck in Cuba for the duration of the war.

Ernest L. Eliel in Havana in 1942.

My family was scattered by that time. My brother in Holland had managed to get out and was living in Brazil. Both my brother in England and my mother (who had been unable to return to Palestine after my father's death in Holland in October 1939 but had managed to join my brother in London) had been interned. By 1941, my mother had been released, and my brother, who was unable to join the British army because he was a foreigner, was a member of the Auxiliary Military Pioneer Corps in Britain, which was responsible for clearing the rubble after air raids. A Dutch friend of my Amsterdam-based brother looked after me in Cuba. He told me, "Wherever you are, act as if you were going to spend the rest of your life there, for if you do not do that, you will uselessly fritter away your time." He followed this advice by asking me what I would do if I were to spend the rest of my life in Cuba. I unhesitatingly answered that I would go back to studying chemistry. "Go to it" was the expected advice he gave me.

I had already received permission to use the library at the University of Havana and to audit various courses. I decided to repeat the first-year courses (which I had nearly completed in Edinburgh) to learn Spanish, which I did not know when I arrived in Havana. With my background in Latin and French, I found Spanish easy to learn.

Matriculating in the University of Havana was a much more difficult problem. My documents from the University of Edinburgh (which had kindly given me credit for the interrupted freshman courses) had to be notarized by a Cuban consul. The Pennycocks obtained duplicates and took them to the consulate in Glasgow. After I received the notarized documents by mail, I had to have them further legalized by the Cuban State Department. Fortunately, at the University of Havana one could matriculate (and take examinations—class attendance was not required) at the end, as well as at the beginning, of the academic year.

My documents from Scotland arrived in time, but when I tried to register in June, I was told that I needed a birth certificate. Fortunately I had taken several copies of my birth certificate from Germany, but as it turned out, they were useless because they were not notarized by a Cuban consul! When I tried to explain to the secretary of the university that Cuba was at war with Germany and therefore there were no Cuban consuls in Germany, he threw me out of his office.

I turned to the refugee association to which I belonged. They, in turn, referred me to their lawyer. In Kafkaesque fashion, the lawyer explained to me that to make the birth certificate legal, I would have to return it to Cologne, have it notarized by the Argentine consul there (who represented Cuban interests during the war), and then have it sent to Buenos Aires, where it would have to be legalized first by the Argentine State Department and then by the Cuban consul. After these steps, the birth certificate had to be returned to Havana for final notarization by the Cuban State Department before it could become a legal document in Cuba. The lawyer also wisely told me that this procedure was unfeasible and recommended instead that my mother should declare before a Cuban consul in London that I was born in Cologne on December 28, 1921, and send me the sworn statement. Then I should have the sworn statement notarized by the State Department in Havana to make it a legal document. This was done, but of course by the time the document arrived (wartime overseas mail was subject to censorship and very slow), the matriculation period was over.

Fortunately there was one last chance not to lose the year. A final registration period was scheduled for September, and it was followed by a second round of final examinations. On the first day of the September matriculation, I appeared at the university secretariat with my university transcript and my mother's statement. To my joy, the secretary accepted the documents, but to my dismay, he now asked me for a certificate of identity. When I inquired what he meant by that, he replied that all the documents I showed him were in the name of Ernst Ludwig Eliel and that he had to make sure I was really that person. This seemed easy to prove, because I had with me my Canadian travel paper with my picture (which the Canadians had issued me after discovering that they—or the British—had lost my passport). But the secretary was not satisfied with the Canadian travel papers; he wanted the certificate of identity, and again, he threw me out.

I went straight to the lawyer who had advised me before and asked him what to do. He said that the certificate of identity required two witnesses who had known me for 2 years and a picture. When I told him that I had been in Cuba for less than a year and that no one in Cuba had known me for 2 years, he was unimpressed and asked me to get the witnesses anyway. I was just about to leave his office when he asked me how old I was. I told him I was 20. My age turned out to be disastrous, because it meant that in Cuba I was still a minor, and Cuban law says that to get a certificate of identity, a minor needs the written permission of one of his parents!

It was evident that this permission could not be secured in time. The lawyer, noticing my dismay, fell into a trance from which he soon emerged with an amazing suggestion. He said that he was bound by the law only if he used the official form for the certificate of identity, but not if he made up his own form. He opined that the secretary of the university would not know the difference! Thus, when I came back after two hours with my two witnesses and the photograph, the lawyer had a very pretty document with ribbons and sealing wax all ready for me. The next day I took this document to the university and was duly matriculated!

Many years later, one of my colleagues once said that I was a bulldog. I think the story of my travails in Cuba (I could tell other similar stories) shows how I came to be one.

Student Life in Cuba

My studies at the University of Havana were most disappointing. The entire chemistry department had four rooms: a general chemistry laboratory, an analytical chemistry laboratory, an organic chemistry laboratory (which was later shared with physical chemistry), and a lecture room. The analytical laboratory was the best; there at least, I learned qualitative and quantitative analyses in much the same way as a student would in the United States. The organic laboratory consisted of a set of rather simple preparations. The physical chemistry laboratory was terrible; the outcome of one of the gas law experiments depended not on the experimental observations but on data gleaned from a handbook. When I tried to explain this fact to the professor, he did not understand that the experimental data cancelled out in the final result.

Fortunately, there was, in Havana, one good pharmaceutical laboratory owned by Angel Vieta, dean of the Medical School, where some research was being undertaken by George Rosenkranz, who later became president of the Syntex Company in Mexico City. Rosenkranz, a very dynamic and highly imaginative individual, and his colleague Stephen Kaufmann, who worked in the same laboratory, were Hungarian refugees who had earned their Ph.D.s in Switzerland with the Nobel laureate Leopold Ruzicka. They allowed me to work in their laboratory, which was an oasis in what was otherwise a chemical desert. Whatever advanced organic chemistry I learned as an undergraduate (other than what I have taught myself), I owe to these two men. This is true, above all, of the laboratory training they provided. Kaufmann was a very thorough person and an excellent experimentalist. By the time I reached the United States several years later, I was undoubtedly better prepared in laboratory techniques than most of my American contemporaries.

As a result of my sojourn in Cuba, I changed my interest from physical chemistry to organic chemistry. In Edinburgh, I had planned to become a physical chemist, and even in Havana, I took some advanced mathematics courses and a course in theoretical physics early on. For theoretical physics, we used the textbook by Lee Page. When we got to Hamiltonian dynamics, the professor said, "From now on, don't ask me what it means," and continued with a purely mathematical presentation that we had to reproduce in detail in examinations. This approach only served to lessen my interest in the subject! In any case, in view of my love for the laboratory, I had never planned to become a theoretician, and the complete absence of people and equipment in the area of experimental physics and physical chemistry made it impossible for me even to find out what those subjects were about. Thus it was easy for Rosenkranz and Kaufmann to turn me into an organic chemist!

For the undergraduate degree (inappropriately called a doctorate in physicochemical sciences), the University of Havana required a thesis. But there were no facilities for thesis work in the chemistry department. Fortunately, because of my connection with Rosenkranz and Kaufmann,

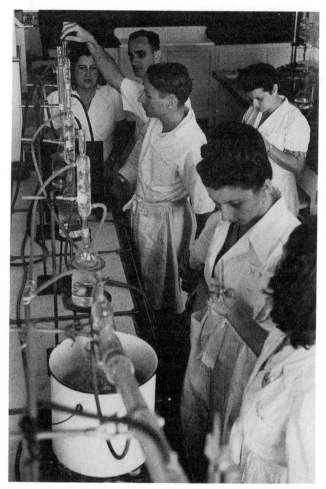

Ernest L. Eliel and other students in a pharmaceutical laboratory in Havana at an extension course in organic analysis and synthesis during the summer of 1945.

I was able to do the thesis work in their laboratory. My undergraduate thesis dealt with a preparation of homoveratric aldehyde (3,4-dimethoxyphenylacetaldehyde) and homoveratrylamine (3,4-dimethoxy-β-phenylethylamine). The work was actually published[2] and constitutes one of my two publications from Havana. The other publication is a paper about chromatography that appeared in the *Journal of Chemical Education*.[3] Chromatography was a technique that I learned in Rosenkranz's laboratory and that fascinated me; the paper, I am afraid, is just a condensation of Zechmeister's book,[4] from which I learned the technique.

I received my degree in May 1946, by which time the war was over and it was no longer difficult to obtain an immigration visa for the United States. By July 1946, I had left what had been my temporary but involuntary home for the previous 5 years for the United States. I arrived with a hundred dollars in my pocket. The United States was still the land of unlimited opportunity!

Academic Career

Graduate Study in Illinois

Before leaving Havana, I had applied for graduate admission to Harvard University, the University of Michigan, and the University of Illinois. Harvard turned me down, with the ironical statement that it was against their policy to admit for a Ph.D. someone who already had one. The refusal from the University of Michigan was more down-to-earth. In 1946 they admitted only residents of the state and returning veterans. Fortunately I was admitted at the University of Illinois, partly because of my academic record and partly because I had done some library work in Havana for a physical chemist by the name of Alfred Lee Sklar, who was a visiting professor in Havana on a Rockefeller fellowship and who was willing to recommend me strongly to Carl S. Marvel. I find it amusing in retrospect that Sklar also told me that if nothing else worked he had such close relations to the University of North Carolina (UNC) at Chapel Hill that he could get me in there. This last-resort possibility did not appeal to me at all, because the chemistry department at UNC was not well known then!

Upon my arrival in the United States, I proceeded directly to the University of Illinois, only to find that before my admission was to be definite I had to find housing in a university town flooded with returning World War II veterans making use of the GI bill, a Congressional act providing for a subsidy for their education. The only way to find a room was to knock on doors and ask for one. The result provided me with an immediate demonstration of the American (or at least Midwestern) kindness and hospitality to a stranger. Almost everyone responded to my request sympathetically, and after a dozen or so calls, I found a room. Subsequently, I was registered as a graduate student.

A month later, I took the usual graduate entrance examinations and, to my surprise, passed all four. Despite the weakness of some of my education in Havana, I did well enough in my course work to be awarded a graduate scholarship at the end of my first two semesters. In fact, at the end of my first semester, I was invited to join a program for the customized synthesis of fine chemicals for industry, which was then extant in the chemistry department of the University of Illinois.

My principal assignment was the preparation of ninhydrin (triketohydrindane). The first step, the Claisen condensation of diethyl phthalate and ethyl acetate in toluene by using sodium sand, was a somewhat hazardous procedure on the five-molar scale I used. In fact, my successors in the job regularly had major fires during this procedure. We were never able to fathom what they did differently from what I did.

Soon came the time to become affiliated with a research director. I had come to Illinois with the idea of working for one of the well-known senior professors: Roger Adams, Carl S. ("Speed") Marvel, Reynold (Bob) Fuson, or Harold R. Snyder. That I might have associated myself with one of the junior faculty—among them Nelson Leonard— never entered my head. This attitude was then, and probably still is, quite prevalent among not-so-sophisticated graduate students, who do not realize that the youngest faculty members frequently have the most exciting problems—because they must still prove themselves—and that they tend to give the most attention to their students.

Because my admission to the University of Illinois was in part due to a recommendation to Marvel (*vide supra*), it might have been expected that I would work for him. However, under the impact of the World War II Rubber Reserve Project, Marvel's research had entirely turned to polymers, an area that did not fascinate me. Adams (the "Chief," who had personally registered me for my courses) was spending much time in Germany and Japan as part of the postwar effort to ascertain what had gone on scientifically in the Axis countries and might not have given me the support that I believed (perhaps mistakenly) I needed. The logical choice was therefore to work with Snyder, especially in the light of indications that Fuson's group was full.

The choice proved to be felicitous. Harold Snyder was a benevolent research supervisor who gave me as much freedom in research as I could use, and both he and especially his wife Mary Jane showed me great kindness at a time when I was still a stranger in a new homeland. Snyder also taught me how to write a scientific paper clearly, concisely, and in good English, a lesson that has proved of enormous value in my subsequent career!

Graduate Research Work

My research problem was to investigate the scope of carbon alkylation with Mannich bases of the indole series. Snyder had then recently discovered[5] (contemporaneously with a group at the Stirling–Winthrop laboratories[6]) that gramine, which can be readily prepared from indole, can alkylate acetamidomalonic ester to give ultimately the amino acid tryptophan (Scheme 1; R = H).

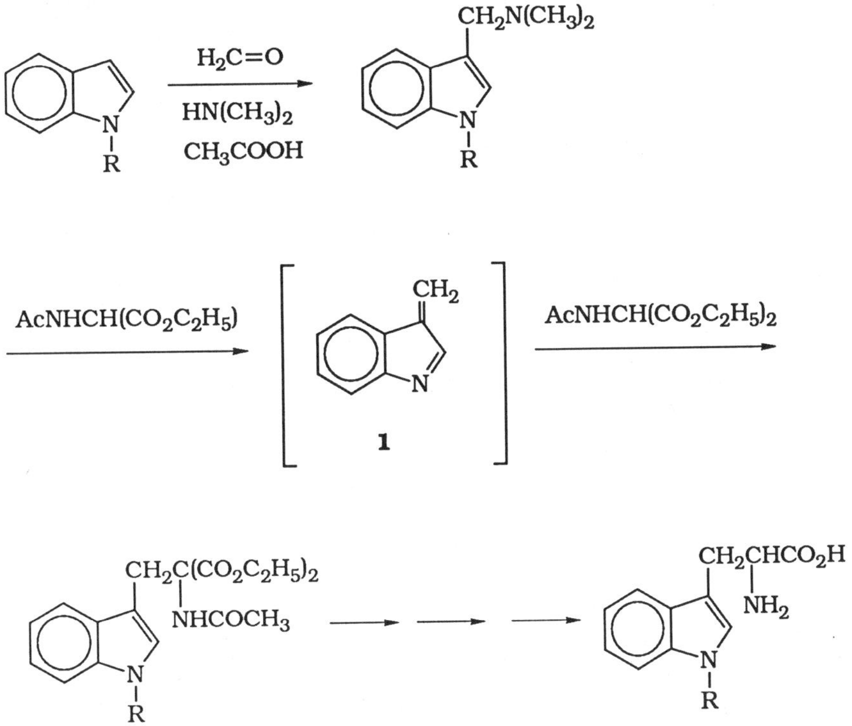

Scheme 1

The reaction was believed to proceed via the methyleneindolen-ine intermediate (1), which then underwent a Michael addition. I was to find out if the reaction would proceed if formation of this intermedi-ate was blocked by making R = CH_3. The Mannich reaction of N-methylindole with formaldehyde and dimethylamine yielded N-methylgramine, but, as expected, N-methylgramine did not react with nucleophiles. However, the absence of reaction could be overcome by substituting the quaternary methiodide of N-methylgramine for the free base, that is, by using the Robinson technique instead of the Mannich alkylation technique. The first experiment involved alkylation with cyanide and yielded 1-methylindole-3-acetic acid after hydrolysis (Scheme 2; 2). Like Stephen Kaufmann during my earlier chemical up-bringing, Harold Snyder insisted on excellence in experimentation, a lesson that I hope and trust I have learned for the rest of my career. Thus I had to synthesize an authentic sample of **2** before we published our first collaborative paper.[7]

To our surprise, the reaction (Scheme 2) produced not only crystalline 1-methylindole-3-acetonitrile but also a second crystalline product, which was separated from the first product by vacuum distillation.[8] I had just read about the S_N2' reaction and speculated that the second product was 1,3-dimethyl-2-cyanoindole. However, Snyder hypothesized that the second product was 1-methylindole-2-acetonitrile, which could be formed from 1-methyl-2-dimethylaminomethylindole, a putative byproduct of the Mannich reaction and a possible contaminant of my liquid N-methylgramine starting material. In 1946, it was difficult to ascertain the purity of liquid substances.

How could I get a lead before synthesizing an authentic specimen of one or the other of the two possible candidates? Snyder told me that another research group had a brand-new instrument, a single-beam infrared spectrophotometer, that might possibly solve our problem. I contacted a Mrs. Johnson, who operated the instrument, and was told that if I gave her both nitriles, she might possibly tell if one was conjugated. Indeed one nitrile was conjugated, as evidenced by its lower frequency C≡N stretch. The subsequent synthesis of the authentic hydrolysis product, 1,3-dimethylindole-2-carboxylic acid, showed that my hypothesis was correct. However, to this day, it is not certain that the reaction was S_N2' and not S_N1', because no kinetic study was performed. Intermediate **3**, which is formed by the loss of trimethylamine, might have been involved.

I soon synthesized 1-methyltryptophan[9] and performed a number of other alkylation reactions with 1-methylgramine methiodide. In 1948, after only 2 years of graduate work, I received my Ph.D. A total of eight publications resulted from my thesis work.

One reason that I was able to work so fast was James H. Brewster, a fellow graduate student in Snyder's research group. Jim Brewster worked on a problem parallel to mine (alkylation with simple quaternary ammonium salts) and, unlike me, was an avid reader and searcher of the chemical literature. He did most of my literature searching for me (in the process of doing his own) and thereby freed my time for the laboratory work that I loved. Both Jim and I were first exposed to stereochemistry in a graduate course taught by Nelson Leonard. Brewster made extensive use of stereochemical principles in his research, and he kindled my own interest in the subject. Later, I had the pleasure of collaborating with him on a chapter in the series *Organic Reactions* dealing with alkylations with amines and quaternary ammonium salts.[10]

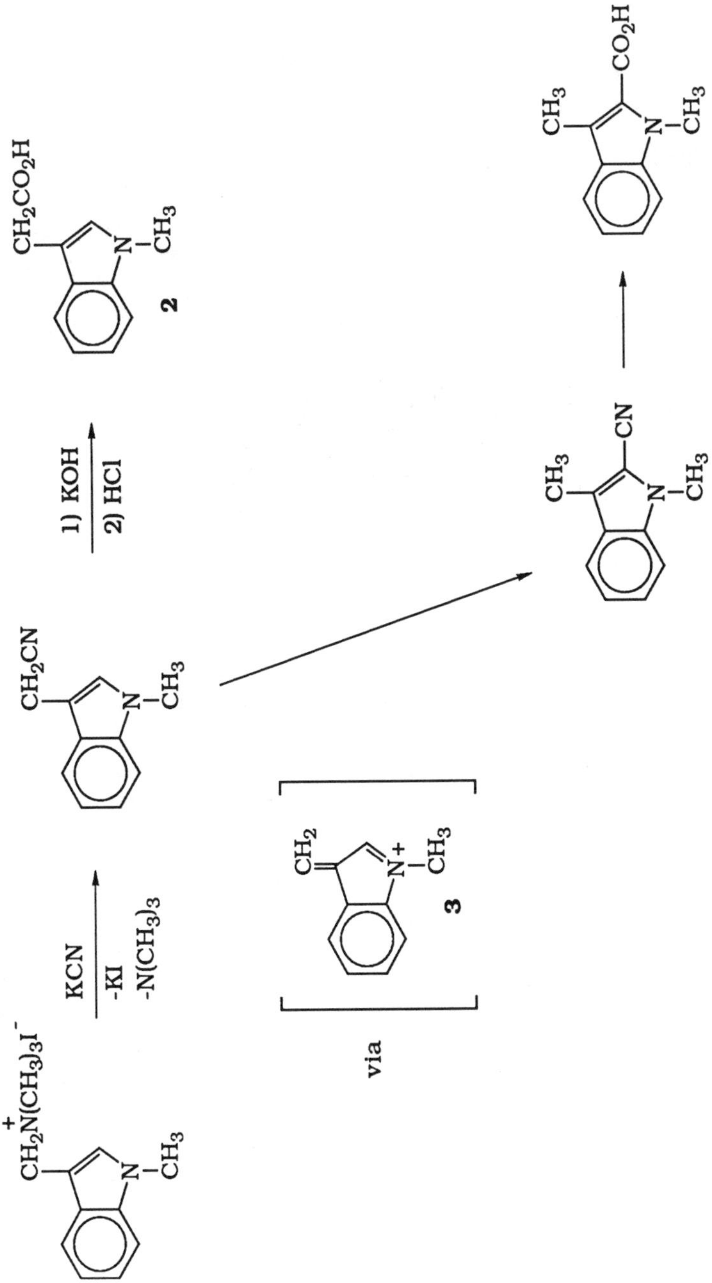

Scheme 2

Ernest L. Eliel (far right) in Illinois around 1947, with fellow graduate students.

Another fellow graduate student in Snyder's group was Dick Heckert, who was recently chairman of the board and chief executive officer of E. I. du Pont de Nemours and Company. After hardly seeing each other for many years, we have recently had the opportunity of working together on a fund drive for the American Chemical Society (ACS).

Search for an Academic Position

I must report here a brief but important meeting that occurred, I believe, in late 1947. Ludwig F. Meyer, one of my parents' closest friends, asked me to meet him for lunch in Chicago, where he was attending a professional meeting. Meyer had been a highly renowned pediatrician and a professor of pediatrics at the university in Berlin until the Nazis dismissed him from his position and forced him to emigrate; he spent the rest of his life in Tel Aviv, Israel. He wanted to see me to discuss with me my plans for the future.

I told Meyer that I wanted to pursue an academic career but that I was concerned as to whether I had the plethora of good and original ideas that a successful researcher must surely have. His answer, which I

At the symposium held on Harold Snyder's retirement at the University of Illinois in 1976. Left to right: Ernest L. Eliel, Harold Moore, Harold Snyder, Richard Heckert, James Brewster, and Karl Folkers.

remember clearly to this day, was, "You need to have some good and original ideas, but it is equally important that you stay with the ideas until they are successfully executed and the results are published." I have taken this advice to heart.

Indeed, in anticipation of finding an academic position, I was beginning to think of new lines of research. Snyder had warned me not to continue my Ph.D. project once I was on my own. In fact, I did work on a few potboilers related to my Ph.D. work for about 2 or 3 years after I received my doctorate. These activities resulted in several undistinguished publications.

At that time, there was a young instructor at the University of Illinois by the name of Elliot Alexander. His field, physical organic chemistry and reaction mechanisms, apparently was not held in high regard at Illinois in the late 1940s. Alexander taught undergraduates only during the regular academic year, but in the evenings, he presided over an eager group of graduate students speculating about the mechanisms of reactions chronicled in Morton's book *The Chemistry of Heterocyclic Compounds*. In the summer of 1947 (which comprised a full semester because the wartime three-semester-per-year arrangement was still in effect), Alexander was assigned to conduct a seminar course in organic reaction mechanisms. It was for this course that the registered graduate students dug up the material that was later organized into his book *Principles of Ionic Organic Reactions*, which was published in 1950.

The final examination was a set of 30 questions formulated as follows: "Predict, on mechanistic grounds, the following reactions (starting materials given)." It was rumored that the best paper had fewer than 20 correct answers and that 12 correct answers were enough for an A and 5 enough for a B. Unfortunately, on the way back from the Chemistry Annex, where the examination had been given, Alexander ran into Speed Marvel, who prided himself on not knowing anything about reaction mechanisms. Marvel looked at the examination and is said to have answered 29 of the 30 questions correctly, on the basis, of course, of his chemical knowledge and intuition and long-time experience that we graduate students did not have. The incident is said to have been quite a blow to Alexander's desire to teach a specific course in reaction mechanisms. Alexander sadly perished 3 years later, when his private airplane crashed.

Alexander was interested in obtaining optically active RCHDR' compounds, a task that had eluded Roger Adams and others years earlier.[11] He subsequently succeeded in obtaining optically active deuterated menthanes, but the work was received with some skepticism, because the approach (catalytic deuteration of 2-menthene) was mechanistically unclear and the product was a liquid. Liquids were difficult to purify, and the possibility that the optical activity was due to impurities could not be excluded unequivocally. I resolved to tackle this problem in what I thought was a convincing approach.

But first I had to find an academic position. In 1948, the search for an academic appointment was far from trivial, because many of the large postwar crop of Ph.D.s had the same idea. I applied for a postdoctoral position at the Massachusetts Institute of Technology (MIT; with A. C. Cope) and for a postdoctoral fellowship at Illinois. But when Charles C. Price (who had just recently left a professorship at Illinois to become head of the chemistry department at the University of Notre Dame) offered me an instructorship in the late spring of 1948, I accepted, largely on the advice of Snyder, who felt that a tenure-track position was preferable to a postdoctoral stint. Thus, in September 1948, I became an instructor at the University of Notre Dame at a salary of $4000 for 12 months, a sum that was not munificent even in 1948 dollars.

I was told that I would have to teach summer school for two summers out of three, that I would be provided with an office–laboratory and routine chemicals, and that it would probably take 2 years before I could be admitted to the graduate faculty and thus be allowed to supervise graduate students. However, I was allowed to work with senior students.

As Struggling Instructor and Assistant Professor

In September 1948, while also preparing for an organic chemistry lecture and laboratory class, I started research with two senior students, Paul Peckham, a premedical student, and Albert Burgstahler, a chemistry major who is now on the faculty of the University of Kansas. Peckham worked on a potboiler,[12] which, however, provided him with good laboratory experience. Burgstahler started on a synthesis of yohimbine. He was an extremely hard-working and highly motivated worker; his B.S. thesis would have been accepted for an M.S. in most universities and for a Ph.D. in some. Several publications resulted from his work.

I myself embarked on the synthesis of nonracemic $C_6H_5CHDCH_3$, with the feeling that if the product showed measurable optical activity, purification of a solid derivative would demonstrate beyond reasonable doubt that the activity was due to the deuterated material and not to an impurity. The work as eventually realized[13] is shown in Scheme 3.

The reduction of halides with $LiAlH_4$ (LAH) or a LAH–LiH combination had just been published.[14] There was evidence that the reaction proceeded with inversion of configuration. The known resolution of the alcohol via the brucine salt of the phthalate, although tricky like so many resolutions, was successful, as was the subsequent conversion

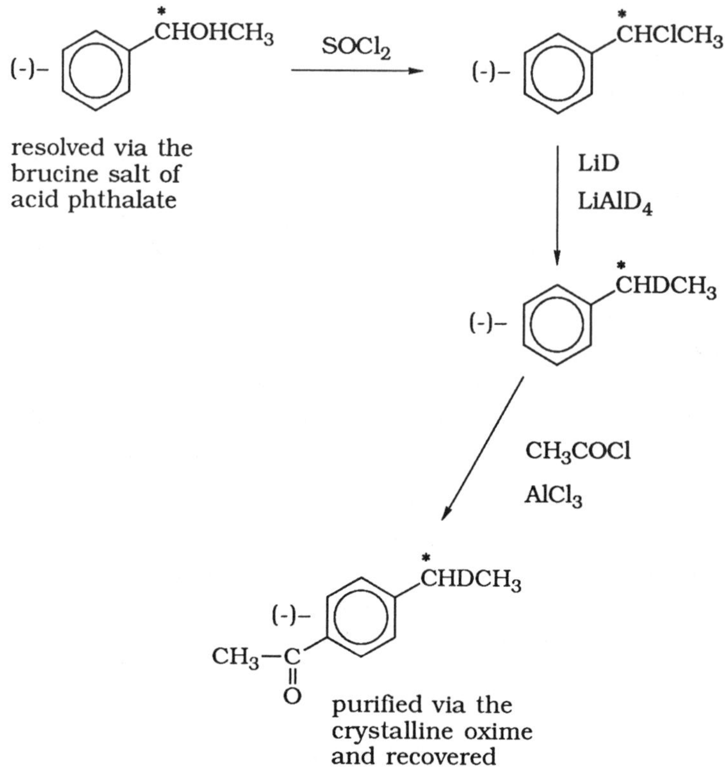

Scheme 3

to the chloride, which, however, involves partial racemization. To carry out the reduction, I required 5 g of $LiAlD_4$ at \$14/g and 10 g of LiD at \$7/g. These were nontrivial sums at that time.

  The approval of Charlie Price, the department head, was needed for the expenditure of \$140. "That is a lot of money—what do you want to do with it?" he asked. I proceeded to give him a five-minute research proposal on the blackboard in his office. Not only did he allot me the money but he also modified my proposal for the better. I had planned to nitrate, reduce, and acetylate the deuterated ethylbenzene to the *para*-substituted acetamide. Price pointed out that the nitration might lead to H–D exchange, racemization, or both, and in any case, my proposed reaction sequence would give an undesirable *ortho–para* mixture. Altogether, Price—a brilliant and highly inventive scientist—gave me much helpful scientific advice during the 6 years that we were together at Notre Dame. To some extent, his mentorship compensated for my lack of postdoctoral experience.

The success of the project on deuterated ethylbenzene gave the needed boost to the start of my research career. I learned that my work was favorably mentioned even in Paul D. Bartlett's course on reaction mechanisms at Harvard and my results evidently provided impetus for important work elsewhere.[15]

However, rather than carry on with applications of RCHDR' compounds, which others did quite successfully,[15] I decided to concentrate on elucidating the mechanism of the hydride reduction of halides. At least in the short run, this project might not have been the best one to pick; in any case, the problem turned out to be quite difficult.

Ordinary tertiary halides, which would be required for the classical stereochemical study involving reduction to RR'R''CH, were known[14] not to undergo S_N2 reductions with LAH. I decided to study an activated halide, $C_6H_5(CH_3)CClCO_2H$, with the (probably vain) hope that reduction of the halide might precede reduction of the carboxylate. The problem proved tractable and publishable[16] in the hands of Jerry Freeman, an able senior student, who is now on the faculty of the University of Notre Dame. However, the interpretation of the results was not immediately obvious because of the presence of two reducible groups and our lack of knowledge then as to which group was reduced first (see below).

{In December 1949, I was married to Eva Schwarz, whom I had met in New York City (where I attended the bar mitzvah of one of my nephews, on a date arranged by that nephew's grandmother). Like myself, she came from Germany, but by the time we were married, she had lived in the United States for 12 years. She was prepared to forgo a promising career as a journalist at *Time* magazine to join me in a small midwestern town. A good bit of my subsequent success as a scientist, I am sure, is due to her unfailing support and to her willingness to take care of our household and, later, of our two children while I spent my time in the laboratory or at the writing desk.}

As a Member of the Graduate Faculty

In 1950 I became a member of the graduate faculty. The late Donald Rivard was my first graduate student. I put him on the yohimbine synthesis, which, however, never panned out, perhaps in part because of my inexperience in synthetic chemistry. He was soon followed by Sister Carolyn Herrman, Richard McBride, and James Traxler. Sister Carolyn and J. Traxler (and later Tom Prosser and David Delmonte) got the hydride problem back on track[17] by showing that the reduction of haloacids and also of their esters involved a halohydrin–hydride com-

Eva and Ernest at their engagement in the summer of 1949 on Fire Island.

plex $[XCRR'(CH_2)_nO]_xAlH_{4-x}$, which would then undergo facile intramolecular reduction when $n = 0$ or 1.

Dick McBride worked on a problem that had been started in Havana and that was continued for a while, with Snyder's blessing, at Illinois, namely, the reaction of phenylacetaldehydes with ammonia to give amines. McBride showed[18] that the reaction was an abnormal Chichibabin reaction (Scheme 4). Similar reactions had already been recognized by Chichibabin himself in the 1920s.

Several things happened on an intellectual plane during the years 1950–1952 that decisively shaped my later career. After getting his B.S. in 1949, Burgstahler proceeded to Harvard for graduate work. There he came in contact with D. H. R. Barton, who was then spending a year at Harvard. In connection with the proposed yohimbine synthesis, Burgstahler had studied earlier degradative work on the alkaloid and its stereoisomers; some of this work had been mechanistically obscure. Now, in a series of letters inspired by Barton's lectures,

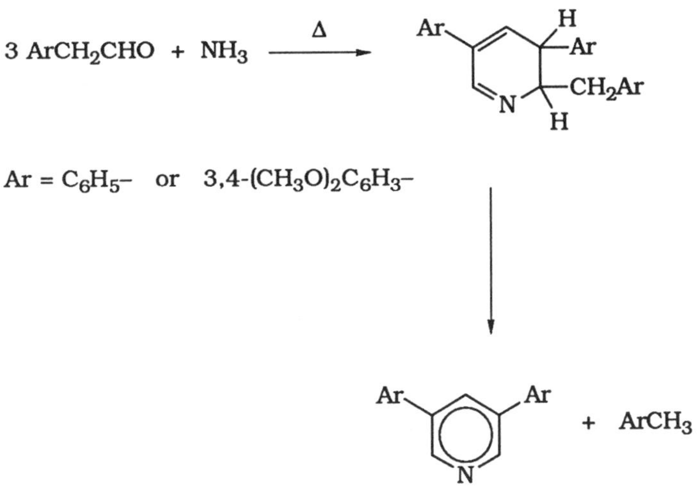

$$3 \text{ ArCH}_2\text{CHO} + \text{NH}_3 \xrightarrow{\Delta}$$

Ar = C_6H_5- or $3,4\text{-(CH}_3\text{O)}_2\text{C}_6\text{H}_3-$

+ $ArCH_3$

Scheme 4

Burgstahler logically interpreted all those reactions in conformational terms. Barton's pioneering paper on conformational analysis[19] appeared soon thereafter, and in early 1950, Barton himself came to Notre Dame to give a lecture, which I found electrifying and which set me to think about conformational analysis in earnest.

Notre Dame has an endowed lectureship series called the Reilly Lectures. In the 1940s and 1950s, Reilly lecturers came for periods of several weeks. In the spring of 1950, the lecturer was Vlado Prelog, who even then was vitally interested in stereochemistry, as a result of his classical work on quinine configurations. We developed a close personal relationship. It was Prelog's first visit to the United States, and the fact that Eva and I were fellow Europeans and spoke German fluently created a special bond. Prelog told Eva at the time that he put people in three categories: those who like Bach and smoke a pipe (++), those who like Bach and do not smoke a pipe (+), and those who like Wagner (−). Eva liked Wagner, but she also liked Bach, and so she just passed muster. She did not smoke a pipe! Prelog's visit not only created a lasting friendship but also contributed considerably to my interest in and understanding of stereochemistry. The 1952 paper[16] on the hydride reduction of (−)-$C_6H_5C(CH_3)ClCO_2H$ reflects, in part, this new understanding.

In 1951 the Reilly lecturer was Michael Dewar, who was then a young and eager physical organic chemist temporarily housed at the Laboratories of Courtaulds, Ltd., in England. I remember a vigorous discussion we had one evening, in the course of which we were thrown out of a restaurant at midnight (because it was closing), and we adjourned to a diner until 2 a.m. Dewar was of considerable help to me in defining the problem of the hydride reduction of halohydrins. During his stay at Notre Dame, Dewar also wrote a series of involved papers that were published in the *Journal of the American Society* in 1951. Later, Dewar claimed that these papers had anticipated the Woodward–Hoffmann rules.

The summer of 1951 was my third summer at Notre Dame, but the promised 3-month leave did not come about because the department was short of faculty for the summer session, which was then very popular among high-school teachers. But I did finally get away in the summer of 1952. By chance, Melvin S. Newman invited me to spend the 3 months at Ohio State University (OSU), where there was to be a high-level summer course on stereochemistry and reaction mechanism. The lecturers, as I now recall, were to be David Curtin, Jack Roberts, Don Cram, Herbert Brown, and Robert Taft. In addition, Newman offered me laboratory space so that I could continue my hydride work.

The availability of laboratory space proved to be very important, because I was then analyzing the reduction of $CH_3CH_2CHBrCH_2OH$ and $(CH_3)_2CBrCH_2OH$ with $LiAlD_4$. Harold Shechter at OSU had a mass spectrometer that I was allowed to use toward that end. I became very interested in the mass spectrometry of organic compounds when I found that $(CH_3)_2CDCH_2OH$ (one of the possible reduction products of $(CH_3)_2CBrCH_2OH$) gave not only $(CH_3)_2CD^+$ ($m/e = 44$) and CH_2OH^+ ($m/e = 31$) but also $(CH_3)_2CH^+$ ($m/e = 43$) and $CDHOH^+$ ($m/e = 32$). Clearly, rearrangements were occurring in the mass spectrometer, and therefore, authentic labeled species had to be prepared for the product analysis. The possibility of rearrangements in the spectrometer also raised important questions concerning the then largely unknown mass spectral breakdown of functionalized organic compounds. Hydrocarbons had been extensively studied in the petroleum industry, but largely for purposes of quantitative analysis, because gas chromatography was then only just being invented.

When I returned to Notre Dame in the fall, I found that a mass spectrometer had just been bought by the Radiation Project, a collaborative research project financed by the Atomic Energy Commission (AEC). I had hoped to use it to continue my studies, but I was informed by the two physical chemists in charge that I could not use the spectrometer because I would probably contaminate it with organic compounds of low volatility! We did only limited work in the area thereafter.[20,21]

The reader might recall that this was in 1952, several years before the pioneering work on mass spectrometry of organic compounds was undertaken by Klaus Biemann, Carl Djerassi, and others. However, this was not the only promising research problem that foundered because of external circumstances. As a result of some work we did around 1955,[22]

I became interested in displacement reactions in polar aprotic solvents. The project died because I could never interest a student in it, perhaps because I had not, myself, defined and elaborated the objectives carefully enough.

Some Disappointments

The academic year 1952–1953 may have marked the nadir in my scientific career. Despite great diligence, I had authored or coauthored only seven independent papers in 4 years, and some of these papers were potboilers. In 1950 I had been promoted, with some reluctance, from instructor to assistant professor. My teaching tended to be over the heads of the undergraduates. During my first year, I almost ruined the career of the top physics major at Notre Dame by trying to give him a C in organic chemistry. From then on I was excluded from teaching the undergraduate organic chemistry lecture courses for a number of years.

Now the matter of tenure was coming up. My elder daughter, Ruth, was born in February 1953, and my wife, Eva, had to give up the secretarial job that had helped to supplement my meager salary. The situation in chemistry at Notre Dame had deteriorated. Of the four senior organic chemistry professors, Kenneth Campbell had resigned to become research director at Mead–Johnson and Charles Price had stepped down from the headship after running unsuccessfully for senator from Indiana in 1952 (later, in 1954, he also ran for the House of Representatives). By 1954 Price, too, left Notre Dame for the University of Pennsylvania. Only Lee Benton and George Hennion, the acetylene chemist, remained. They faithfully continued to be the kind of senior mentors whom every young faculty member needs. I believe it was Hennion who persuaded me not to give the C to the brilliant physics major. I remember him telling me this axiom about Notre Dame when I first came: "Don't ever forget that the Community of the Holy Cross owns this place, lock, stock, and barrel."

I tried for a position at a couple of the prestigious universities in the Northeast but did not succeed. My record was not adequate. But at least I had not lost faith in myself. My publication record improved somewhat in 1953, but I believe it was largely the abiding faith in me of the late Milton Burton, a radiation chemist who was then the director of the Radiation Project at Notre Dame, that was responsible for my being given tenure in 1953. I am glad that I did not disappoint this warm and kind-hearted man, who later became a good personal friend, despite the fact that we often differed on both scientific and administrative matters.

Important New Ideas

Conformational Analysis of Mobile Systems

During the summer at OSU in 1952, I had been thinking about the conformational analysis of mobile systems, such as nucleophilic displacements on cyclohexyl bromide. Cyclohexyl bromide was known from Hassel's earlier work[23] to exist as a mixture of equatorial and axial conformers. Barton's initial work had dealt virtually exclusively with conformationally biased or rigid systems, because such systems most clearly and simply demonstrate the conformational principles of stability and reactivity discussed in his original paper.[19] The chemical behavior of conformationally mobile systems, such as cyclohexyl bromide, was not understood during the early 1950s.

In this connection, I was intrigued by an article by Read[24] concerning the nitrobenzoylation of the four diastereomeric menthols and especially by the fact that neoisomenthol (Scheme 5; 4) reacted faster than neomenthol (5). By this time Barton's lecture and my correspondence with Burgstahler had taught me to think of cyclohexanes as chairs, and my interpretation of Read's results was that neoisomenthol would react via the conformation 4b, with equatorial hydroxyl, rather

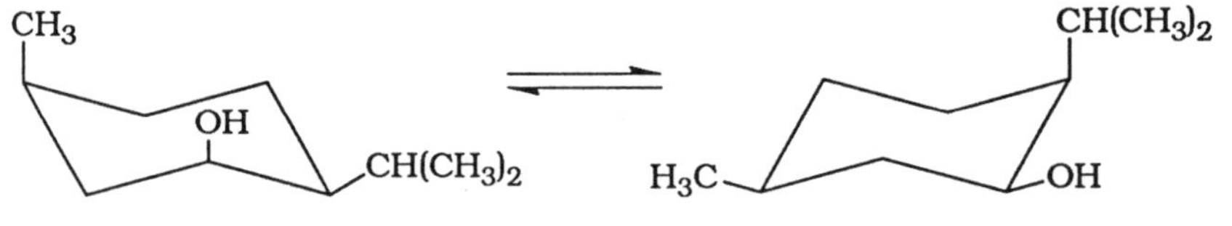

4a   4b

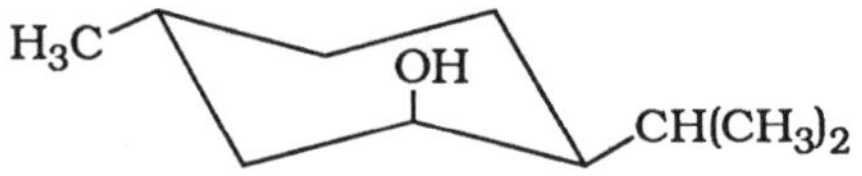

5

Scheme 5

than **4a**, with equatorial isopropyl, inasmuch as **4a**, with its *syn*-axial methyl, should react more slowly than neomenthol (**5**), which had an axial OH but not a *syn*-axial methyl. [The argument was put somewhat differently in the original paper,[25a] because I did not realize in 1952 that neomenthol is conformationally homogeneous (Scheme 5; **5**). This homogeneity was shown much later,[25b] along with a demonstration by NMR spectroscopy that neoisomenthol, indeed, exists partially as conformer **4b**.]

At OSU when I told Herbert Brown about this interpretation, he thought that the hypothesis was very novel, inasmuch as it was then believed that the menthols all existed predominantly with equatorial isopropyl groups and because it was also believed that ground-state conformation was solely responsible for reactivity. Brown suggested that I publish the idea, and with his encouragement (which I needed because my scientific upbringing had conditioned me against publishing *"gedanken* experiments"), I did.[25a] The paper was somewhat embryonic, but it turned out to be a harbinger of much more important things to come. The seminal idea was that a compound might pass through a conformation in the transition state that corresponds to a minor conformation in the ground state; as we shall soon see, this possibility leads to important and far-reaching consequences.

My contact during that summer with Brown and Newman bore fruit in other ways. After the summer course, Newman organized his later book *Steric Effects in Organic Chemistry* (published by Wiley in 1956). I volunteered to write a chapter on conformational analysis, a topic about which I had become well informed by reading Barton's papers and other related ones. However, I learned that Bill Dauben and Kenneth Pitzer (both at the University of California, Berkeley) had already been chosen for that task. As it turned out, H. C. Brown, who had been the designated author for the chapter on nucleophilic displacement, had to back out because of other commitments and recommended me for his replacement. Thus I came to be part of an exciting and stimulating intellectual task.[26]

Specific Rates of Conformationally Heterogeneous Systems

The intellectual stimulation of the years 1950–1952 finally seemed to make up for my missing postdoctoral education. By 1954, the hydride work was firmly underway, and I began to think more clearly about conformationally mobile (or heterogeneous) systems. What would be the specific rate of reaction of such a system in terms of the hypothetical rate constants of the axial and equatorial conformers? In September 1954, I attended the Reaction Mechanisms Conference in Durham, NH, and in order to recoup some of my travel expenses, I carried as passengers, from New York City to New Hampshire, Nathan Kornblum (of Purdue University), Ronald Bell (then at Oxford University), and Richard Noyes (then at Columbia University). I posed the problem just stated, and we discussed it. None of the eminent physical organic chemists came up with a solution.

During that fall (1954), the solution suddenly occurred to me: The specific rate k for the mobile system is the sum of the specific rates of the contributing conformers weighted by their respective mole fractions n. At that time, I formulated the solution only for a cyclohexyl system with its two conformers, equatorial (e) and axial (a), so that

$$k = n_e k_e + n_a k_a \tag{1}$$

or

$$k = \frac{k_a + K k_e}{K + 1} \tag{2}$$

in which $K = n_e/n_a$ is the conformational equilibrium constant. In fact, I thought at first that the rate of the minor (axial) conformer could be neglected, so that $k = n_e k_e$, but David Curtin (who had then just moved from Columbia University to the University of Illinois) disabused me of this oversimplifying assumption. At the same time, both Curtin and Bill Dauben (to whom I also communicated the solution in a letter) confirmed that my approach was correct. This confirmation was very important, because my solution was in apparent (but not real) contradiction with the Curtin–Hammett principle enunciated that year.[27a] The principle states that the ratio of reaction products (not rates) from a conformationally heterogeneous system is independent of the ground-state populations.

Provided that the rate of interconversion of the starting conformers is rapid, the Curtin–Hammett principle may be stated as follows:

$$r = e^{-\Delta\Delta G^{\ddagger}/RT} \tag{3}$$

in which r is the product ratio, and $\Delta\Delta G^{\ddagger}$ is the difference between the free energies of the transition states leading to the two different products corresponding to the two conformations of the starting material. In this form (equation 3), r, indeed, appears to be independent of ΔG°, the energy difference between the two conformations of the ground state, or of K, the corresponding equilibrium constant for the equilibrium between the ground-state conformers. However, it is also true that

$$r = K\frac{k_e}{k_a} \qquad (4)$$

When stated in this form, the product ratio r does seem to depend on K, as well as k_e and k_a. Equation 3 is true, because in a given system, a change in K in equation 4 will produce a compensating change in the ratio k_e/k_a. The resulting potential confusion over the Curtin–Hammett principle has been dispelled finally in a detailed review by J. I. Seeman.[27b]

Research Support

In 1953, after 5 years, I finally received major research support. Up to that time, I had had only one starter grant from Research Corporation, which was given in 1951 and renewed in 1953. Financial support of graduate students was less important then than it is now, and money was quite hard to get. The National Science Foundation (NSF) started in 1950 but did not provide support until a few years later. The Petroleum Research Fund (PRF) started making grants only in 1954. The National Institutes of Health did provide support for health-related projects, and Kenneth Campbell had had money from that source. Some of C. C. Price's support came from industry. Don Rivard, my first graduate student, at first made do with the GI bill, as did Dick McBride and Phil Wilken; Rivard later received a fellowship. Tom Murray, an M.S. candidate, had ROTC (Reserved Officer Training Corps) support, whereas Jim Traxler did get some limited supported from the Research Corporation grant. Sister Carolyn Herrman, like most nuns then, was supported by her order, as was the Reverend Conrad Pillar, my first graduate student to work on a conformational problem. Pillar established that the six-membered ring in cis-oxahydrindane was a flattened chair rather than a boat (Scheme 6).[28] At that time, the fact that boat conformations of saturated six-membered rings are high in energy and therefore rare was not as clearly understood as it is now.

Two things happened in 1953. First, Milton Burton, who was by then receiving major support from the AEC (Atomic Energy Commission) for the Radiation Project, which had been underwritten initially

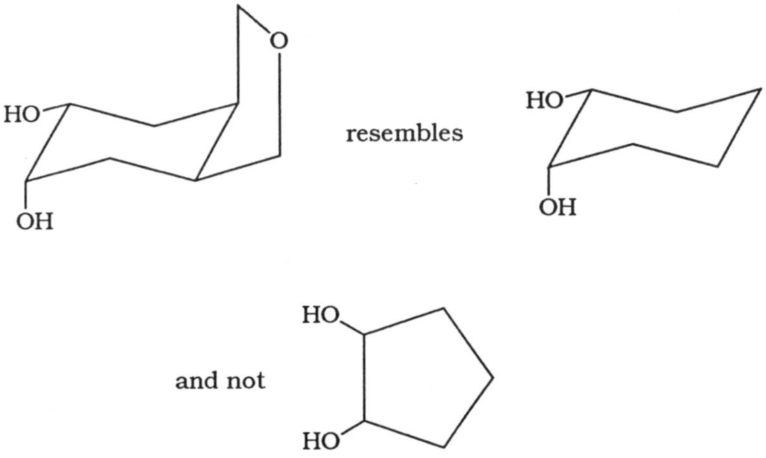

Scheme 6

by the Office of Naval Research, offered to make me a member of that project. Burton was willing to support the hydride work if I would also initiate some work on free-radical reactions related to radiation chemistry. Over the next 10 years I worked on free-radical coupling reactions and various types of free-radical aromatic-substitution processes.

On the strength of the AEC support, I engaged my first postdoctoral collaborator, Fabian Fang, in 1954, and many of my postdoctoral associates during the next 6 or 7 years were underwritten by what later became the Radiation Laboratory. Among them were several who later achieved notable academic or governmental positions: Samuel H. Wilen (of the City University of New York), the late Zoltan Welvart (of the Centre National de la Recherche Scientifique, Gif-sur-Yvette and Thiais, France), Jean-Claude Richer (of the University of Montreal), Osamu Simamura (of the University of Tokyo), Jadu Saha (of the Department of Agriculture, Canada), and others.

Second, in 1953 I applied for and received a sizable grant from the Army Office of Ordnance Research (AOOR), now the Army Research Office, for work on conformational analysis. As a result of this influx of support and, I suspect, also because two of the popular senior research advisers, Campbell and Price, had departed by 1954, four excellent graduate students signed up with me that fall: Carl Lukach and Rolland Ro in the area of conformational analysis and Tom Prosser and David Delmonte in the hydride field. Soon both of these projects were bearing rich fruit, and by 1956, my research career was securely under way.

Equilibrium Method for Conformational Equilibria

However, there was a serious disappointment. In late 1955, Saul Winstein and N. J. Holness at the University of California, Los Angeles (UCLA) published a lengthy paper[29] in which equation 1 was anticipated and applied. Even though we had derived the equation in a different way and applied it to different cases,[30,31] we were scooped. The experience was so much more painful because I had had an exchange of information with Winstein when he was Reilly lecturer at Notre Dame in the summer of 1953. I thought then that he was not working on conformationally heterogeneous systems. (The detailed circumstances of this matter have been chronicled elsewhere.[32] Winstein's plans had evidently changed in the ordinary course of events between 1953 and early 1955, when his paper was submitted, as so often happens in research.) In any case, Winstein's work helped us in suggesting the *tert*-butyl group as a conformation-holding group. We had planned to use less-effective methyl groups. Although the preliminary communication in *Chemistry & Industry* (then still a favorite vehicle for such communications) met no problems, my magnum opus, in the form of three manuscripts, encountered rough going at the hands of the referees when it was submitted in early 1957. The problem with the referees happened despite the fact that the second paper[33] embodied an entirely new method, the "equilibrium method" (Scheme 7) for determining conformational equilibria quantitatively.

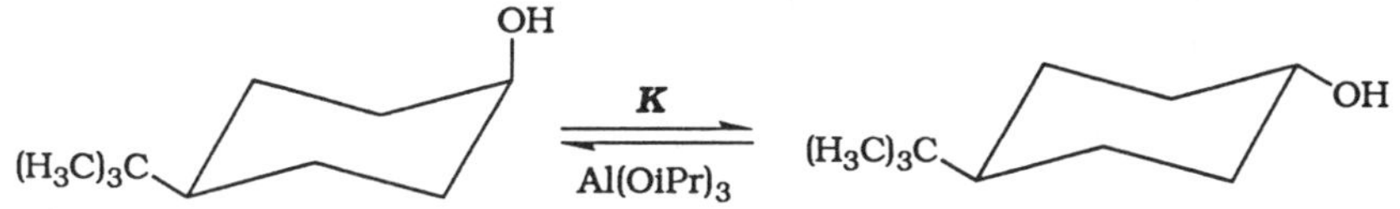

Scheme 7

The reagent used in 1956 was aluminum isopropoxide in the presence of a small amount of ketone. Later we used Raney nickel for this equilibration (*see* Scheme 14). This method turned out to be much better than that based on a regrouping of equation 2:

$$K = \frac{k_a - k}{k - k_e} \tag{5}$$

in which k_a and k_e are specific rates, respectively, for purely axially and equatorially substituted cyclohexanes (usually *cis-* and *trans-4-tert-*

butylcyclohexyl-X compounds), and K is the equilibrium constant for the equilibrium shown in Scheme 8.

In the equilibrium method, it is assumed that in one and the same solvent the equilibria shown in Schemes 7 and 8 are the same. This assumption has turned out to be much safer than the correspond-

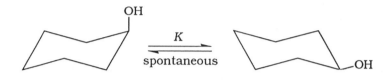

Scheme 8

ing assumption for the kinetic method, namely that the *tert*-butyl "holding" group does not affect the rate of reaction of the functional group X that is being studied (other than by fixing its conformation).

We were not the first to study equilibrations in cyclohexanes; in fact one of the tenets of Barton's classical paper,[19] which was based on information in the older literature, is that equatorially substituted cyclohexanes are more stable than axially substituted ones. Our contribution was to turn this principle into a quantitative method for determining K (Scheme 8) by using the *tert*-butyl conformation-holding group (Scheme 7) first used by Winstein[29] in a different context.

In the first version of the 1957 manuscripts,[30b,31,33] I had not mentioned the 1953 *Experientia* paper[25a] because I considered it too embryonic to be worth citing. But because both referees referred to Winstein's priority in their objections (with one of them complaining that the citation of the Winstein–Holness paper[29] was only reference 12 in the first manuscript of the three!), I did prominently cite my own 1953 paper in the revised version so as to document that we had thought about the problem of mobile systems well before the appearance of the Winstein–Holness publication. The papers were eventually accepted even though one of the referees was still not happy.

I should interject here that this experience is not atypical and was to be repeated several times more in my career in the following sense: When a paper (or sometimes a research proposal) embodies a particularly new idea, the referees often do not seem to understand, and as a result, do not appreciate the idea. However, because they do not understand the new idea, they are unable to shoot it down, and consequently, they concentrate on side issues. I recall that with the paper just mentioned one such side issue was that because we had to prepare and investigate the kinetics of a large number of compounds, we had carried out rate measurements at a single temperature only. Indeed, it would have been preferable to carry out measurements at several temperatures so as to determine activation enthalpies and entropies, but this was not essential to the purpose at hand (*see* equation 5, which should be true at any temperature). In fact, rate measurements at various temperatures might have confused the issue by needlessly complicating it.

I felt that Winstein was the referee who had so objected to the paper, and when I next saw him at the Organic Symposium at the University of Rochester in the summer of 1957, I complained to him that he had not treated me fairly. He was surprised and taken aback. He had not refereed the papers at all (or so he told me, and I believed him). So much for second-guessing the identities of referees! I should add that in later years Winstein was quite supportive of our work, and except for a thesis with Anita Lewin (now a friend and neighbor) in 1962, he did little more in the area of conformational analysis.

My sensitivity to this matter is perhaps understandable. The work on the reactivity of conformationally mobile systems, which was initiated by the 1953 *Experientia* paper,[25] is probably my most important original contribution to chemistry. The ideas that both conformations of a mobile system must be considered in assessing its reactivity and that the minor conformer is not infrequently the one that leads to most or all of the reaction—as for example the axial conformer in a base-induced (E2) H–X elimination from a cyclohexyl halide—are crucial. These ideas have influenced our understanding not only of chemical kinetics but also of enzymatic reactions. In enzymatic reactions, neither the enzyme nor the substrate needs to be in the most stable ground-state conformation in the enzyme–substrate complex. The same argument applies in pharmacology, in the interaction of a pharmacophore with its receptor site.

A Look at Other Places

Industry

Meanwhile, in the summer of 1956, our teaching contracts had been changed from 12- to 9-month commitments with the same salary. This generous arrangement allowed us to draw summer salary from research grants (which became increasingly possible about that time), teach summer school, or do something else. I did not yet have a grant large enough to cover my summer salary, and I greatly dislike teaching summer school (I have not done so in 30 years). So I decided to accept an offer of a summer research position with Standard Oil of Indiana in their research laboratories at Whiting, IN. (Today the company is called Amoco, and their laboratories are in Naperville, IL.)

Despite the objections raised by a friend and colleague (who felt, with some justification, that summers should be used for research and writing), there is no question, in retrospect, that my short sojourn in industry was very useful for several reasons. First, I made the acquaintance of several prominent industrial chemists: D. A. McCauley and Art Lien, who worked in the area of Friedel–Crafts reactions (we were also working on this topic[22]), Al Weitkamp in the area of conformational analysis, Paul Rylander and Sy Meyerson in mass spectrometry, Herbert Friedlander in the burgeoning area of Ziegler–Natta polymerization, Ellis Fields in oxidation reactions, and several others.

Second, through my work with Rylander and Meyerson, I became involved again in the topic of mass spectrometry and the breakdown of parent mass ions. A burning question in those days was whether the $C_7H_7^+$ ion formed by loss of a hydrogen atom from the toluene molecule ion $C_6H_5CH_3 \cdot^+$ was, in fact, a benzyl ion or a tropylium ion formed by rearrangement.[34]

But most important, I found out how an industrial laboratory operates at the level of personnel, research reports, patents, and the choice of problems. Because many university professors have no industrial background but send the majority of their Ph.D. students to industry, I believe such summer experience as I had would be widely useful. In later years, I have kept up my contact with industry through various consultantships, but none of these have taught me as much about industrial operations as my intensive 3-month involvement during the summer of 1956.

Sabbatical

By 1957 I had been in the harness for nearly 10 years, and I felt that it was time for a sabbatical. Notre Dame had no sabbatical policy, but fortunately, the NSF, which by then had been allocated appreciable amounts of money by the U.S. Congress, was awarding senior postdoctoral fellowships at a full annual salary, which in my case was somewhat under $7000 at that time. I applied for and received an NSF fellowship and decided to split my sabbatical year between Harvard, to study kinetics with Paul D. Bartlett, and Caltech (California Institute of Technology), to learn something about NMR spectroscopy with J. D. Roberts. (I had originally contemplated spending part of the year with Saul Winstein at UCLA, but because of the earlier mentioned, although imaginary, problem, I had changed my mind.) Because

C. K. Ingold was to be Reilly lecturer in the fall of 1957, I postponed my leave by one semester. In January 1958, I arrived at Harvard with my wife and daughters, aged 3 and 5.

Harvard. My 6-month stay at Harvard University proved immensely stimulating in a variety of ways. In Bartlett's laboratory, I learned improved experimental techniques for our kinetic work: The rather slow reactions of cyclohexyl bromides with thiophenolate had to be performed in sealed, nitrogen-purged ampules to avoid air oxidation of the thiophenol. At the same time, by observing the free-radical studies of Bartlett's group and by presenting an account of our group's free-radical work in the Bartlett–Westheimer research seminar, I came to the conclusion that our work in that field needed to be improved.

Perhaps more importantly, within the first week of my stay, I gained the respect of the late R. B. Woodward. Woodward's research seminar was on Thursday evenings. So on the first Thursday afternoon I was there, I went to Woodward's office and asked his faithful secretary, Doty Dyer, if there would be a seminar that evening and at what hour. She informed me that the seminar would be at the usual hour of 8 o'clock. (This was the nominal starting time. The seminar rarely commenced before 8:30 or 8:40 p.m.) I left the outer office and had barely walked 10 paces when Doty Dyer called me back and told me that Woodward wanted to speak with me. Realizing how difficult it was to see Woodward even in 1958, I was delighted. After some polite enquiry about the adequacy of my lodgings (we had rented a very comfortable furnished house belonging to Kenneth Thimann), he asked me if I was willing to give the seminar that evening. I had to think fast—if I said no, I would be dead forever in Woodward's eyes—and agreed. I spent the rest of the afternoon at home organizing a series of slides on the hydride work, which, fortunately, I had with me.

Woodward's Thursday evening seminars, which I attended regularly, were enormously stimulating events. At around 10 p.m., after a scheduled talk such as mine, Woodward would write problems from the literature or from his own work on the blackboard and invite solutions. Sometimes the problem was that of a product of unknown structure; at other times it was the mechanism of a reaction of known outcome but obscure course. Solutions were presented on the board by members of the audience on the spur of the moment, criticized, improved, criticized again (often Woodward was the most severe critic), and further improved until either Woodward (and hopefully the rest of the audience) felt that a satisfactory solution was at hand or until (more rarely) everyone gave up in disgust.

Woodward tended to become more and more lively as the evening wore on, whereas the audience (usually including several visitors from nearby institutions such as Brandeis and MIT) gradually became sleepy and slowly drifted away. When the audience fell below a critical mass (usually around midnight), Woodward would suggest that we adjourn to a restaurant on Brattle Street for beer and continue to discuss chemistry and chemists. I usually faded and went home by 1 a.m., when Woodward was still going strong. (The next morning he would be back in his office by 8 a.m. He was known to require very little sleep.)

The hydride work had gone well; by gaining a good understanding of halohydrin reductions, we had discovered the interesting reducing properties of lithium aluminum hydride–aluminum chloride combinations,[35] which consisted mainly of AlH_3 or $AlHCl_2$, depending on the proportions of the ingredients. We observed the rather variegated reduction of a typical epoxide with these reagents (Scheme 9), and we did further work on the reduction of ketals and monothioketals (Scheme 10). Woodward was impressed with the presentation, and our subsequent relations, until his untimely death in 1978, were always very cordial.

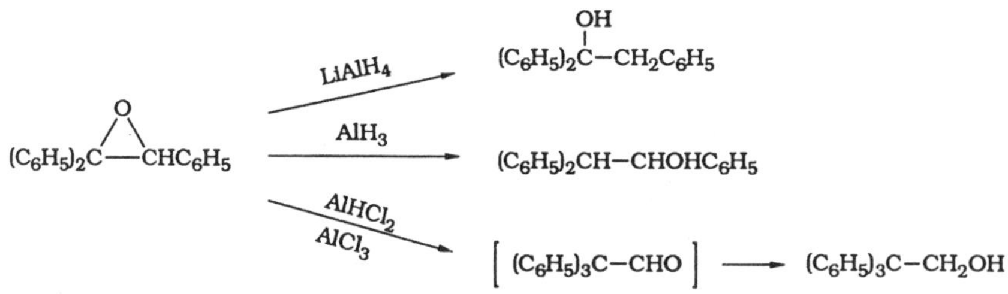

Scheme 9

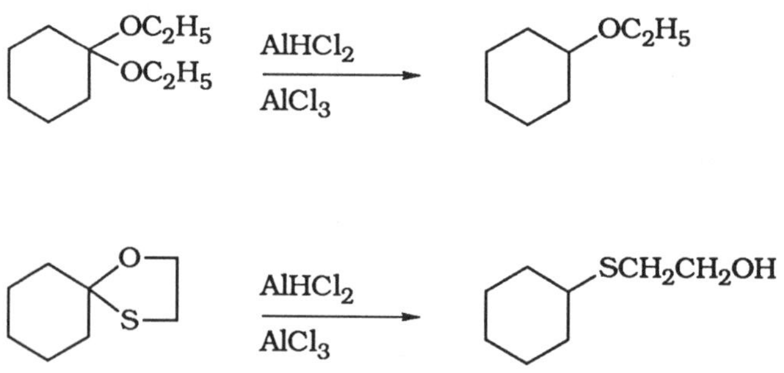

Scheme 10

Lithium aluminum hydride, a reducing agent discovered by R. F. Nystrom and W. G. Brown in 1947, was then, as now, a very important reagent. Because of its novelty and power—it was the first reagent to reduce carboxylic acids to alcohols—it was much in the limelight during the 1950s. Thus modifications of the reagent that change its chemoselectivity attracted considerable interest. In the course of our halohydrin reductions with $LiAlH_4$ (*see* page 25), we had noted the intermediacy of epoxides in some cases, as well as the formation of AlH_3 and aluminochlorohydrides ($AlHCl_2$ and AlH_2Cl, which were first described by Egon Wiberg and later carefully characterized by Gene Ashby). These results led us to the study of epoxide reductions, which are summarized for the case of triphenylethylene oxide in Scheme 9.

While this work was under way, I read a paper about the reductive opening of spirostanes (ketals) by a reagent formed from $LiAlH_4$ and HCl. I recognized that the reagent was probably $AlHCl_2$ or a mixture of $AlHCl_2$ and $AlCl_3$, which thus should be capable of reducing acetals and ketals. The reactions shown in Scheme 10 were discovered as a result of this insight.

One of the very stimulating weekly events at Harvard was the Friday colloquium that featured outside speakers. I myself was invited to give a talk on our work on conformational analysis, an invitation that I greatly appreciated, because until then, I had not had many opportunities to present this work at major institutions or at major meetings, except for ACS meetings. (Ever since Snyder had impressed on me the importance of oral presentations and had twisted my arm to talk about my Ph.D. research at the ACS meeting in St. Louis in the fall of 1948, I have presented papers in at least at one and sometimes two ACS meetings every year). The presentation was very well received, and afterwards, Bartlett remarked that there had been an extra-long round of applause (I had also noticed it), "because," as he put it, "you made the subject so particularly clear." I do not know to this day whether he meant this remark as a true compliment or as a backhanded one (lauding the presentation as opposed to the content).

Caltech. I spent the second half of the sabbatical year at Caltech. I had meant to work with Roberts, but at short notice, Roberts had decided to spend the fall semester of 1958 at Harvard as a visiting lecturer. So instead of my working with him, we rented his house. Despite his absence, the stay at Caltech proved to be very fruitful. The fort was held, so to speak, by Marjorie Caserio, then a temporary instructor (now at the University of California, Irvine), and two postdoctoral researchers: Gideon Fraenkel, now at Ohio State University, and Al Loewenstein, now at the Technion in Haifa, Israel.

The focus of the work was the then-embryonic application of proton NMR spectroscopy to organic systems. Roberts had just finished his 1959 book on the topic,[36] and I was privileged to read it in page proof and to get generous personal help from Caserio, Fraenkel, and Loewenstein when I did not understand a particular point. Moreover, there was a Varian HR-40/60 NMR instrument in Robert's laboratory that I was permitted to use and learned to use.

I decided to investigate the proton NMR spectrum of a conformationally mobile system that I had studied by the kinetic method at Harvard: cyclohexyl bromide. After perusing Gutowsky's papers,[37] I realized that the same equation that we[30,31] and Winstein and Holness[29] had applied to kinetics (equation 1) would also apply to chemical shifts:

$$\delta = n_e \delta_e + n_a \delta_a \qquad (6)$$

in which δ_e is the chemical shift of a given proton (in this case, CHBr) in the equatorial conformer, δ_a is the corresponding shift of the proton in the axial conformer, and δ is the (averaged) chemical shift of the proton in the actual compound (Scheme 11).[38a]

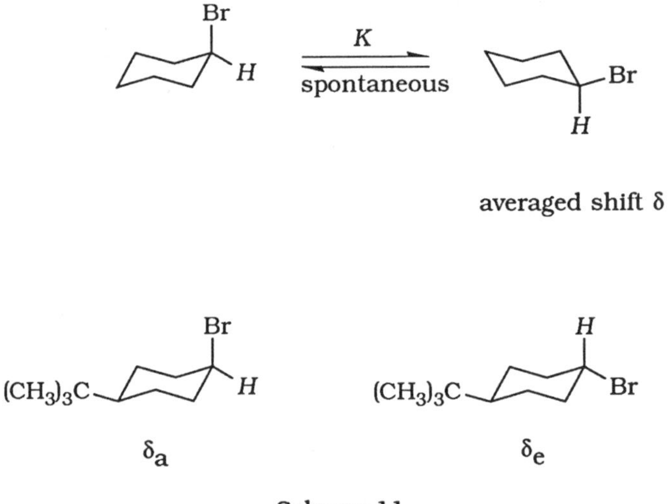

averaged shift δ

Scheme 11

Because δ_a and δ_e for cyclohexyl bromide are not accessible at room temperature (conformational inversion is fast enough to lead to an averaged signal with shift δ), the corresponding shifts in cis- and trans-4-tert-butylcyclohexyl bromides, substances that were already available from our kinetic work, were used instead. Although the innocuousness of the *tert*-butyl holding group is not beyond cavil, this group probably causes less trouble in the NMR spectroscopic studies than it does in the kinetic method. The paper has become a citation classic,[38b] and the ideas it embodies are still used today. That both chemical shifts (equation 6) and coupling constants average in conformationally mobile systems is, of course, part of common knowledge nowadays. But in 1959, this idea was quite novel, and we were the first to exploit it quantitatively to determine conformational equilibrium. Subsequently we determined a number of conformational equilibria in this fashion.[39]

While at Caltech, I repeatedly attended Winstein's Thursday evening seminar at UCLA. Winstein's seminar was quite different from Woodward's. It would start promptly at 7:30 p.m. and usually end at around 10:15 p.m. The major intellectual entertainment was the stimulating but incisive criticism of the speaker (which was not always appreciated!) by Winstein and his postdoctoral associates. Because the criticisms tended to result in frequent interruptions (which were considered entirely appropriate), frequently the speaker (often from UCLA) did not finish his talk and would continue the following Thursday. (Some years later a former Ph.D. student of mine who attended the seminar reported that "The demise of the speaker was postponed to the following week.")

Not unexpectedly, I was asked to give one of the seminars, but unlike my presentation in the Woodward seminar, I did not finish my seminar in one evening. On the way back to Pasadena, one of the passengers in the car, David Schuster (then a graduate student at Caltech and now a professor of chemistry at New York University) asked me, "Is conformational analysis really in such bad shape?" Realizing the wrong impression I had conveyed, I resolved to do better in the following week's continuation, in which I reported on the S_N2–E2 study of the reaction of cyclohexyl bromides with thiophenolate (Scheme 12). This study had been initiated by Ralph Haber (now research direc-

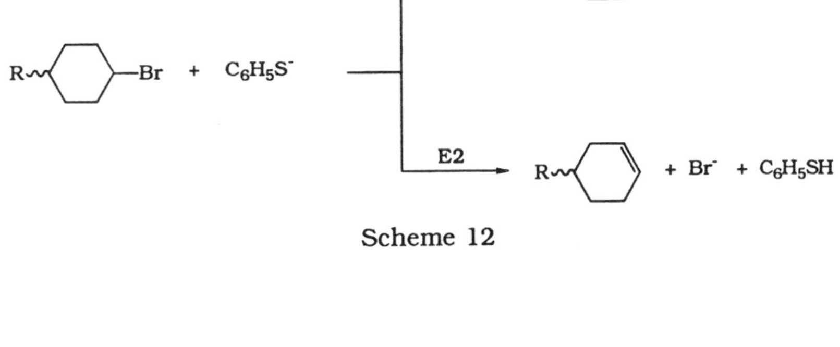

Scheme 12

tor at Abic in Israel) at Notre Dame, and I had completed it at Harvard.[40] The overall rate is determined acidimetrically (because base is consumed), the substitution rate is determined iodimetrically (because thiophenol is consumed), and the elimination rate is determined by difference. Of course it is important to exclude air rigorously, or else thiophenolate is lost by rapid oxidation to diphenyl disulfide.

As I was presenting this work, one of Winstein's collaborators interrupted me and asked if we had done the kinetics by iodimetric titration. When I assented, he said that this procedure had also been tried at UCLA but was found wanting. I asked him if he had worked with sealed ampules, and he responded, "At UCLA we always work with sealed ampules." I thereupon asked if he had flushed his ampules with nitrogen. He looked distressed and shook his head. "At Harvard we always flush our ampules with nitrogen," was my retort. Needless to say I carried the day on this occasion!

My Book and What Followed

My main task at Caltech was to start writing the stereochemistry book for which I had signed a contract with McGraw–Hill the year before. The atmosphere was propitious. Because Roberts was away and George Hammond had joined the Caltech faculty just that fall, things were rather quiet; moreover, I had a comfortable office. Both factors were conducive to writing. The book was based mainly on a course in stereochemistry that I had taught for several years, but of course, I also had to do a great deal of outside reading. On the excellent advice of Kurt Mislow, who was then still at New York University and who was to be one of the two prepublication reviewers (Jerome Berson, then at the University of Southern California, was the other), the book was to contain full references to the original literature, something not common for textbooks and that I had not contemplated at first. I completed more than half of the writing at Caltech; the rest was done at Notre Dame in the summers of 1959 and 1960.

The book, *Stereochemistry of Carbon Compounds*, finally appeared in early 1962. Someone who has never written a book is likely to underestimate the length of the process. Once a complete manuscript has been submitted, it has to be read by the publisher's reviewers, criticized, and revised in the light of the criticism. Then the author must go through the process of obtaining camera-ready drawings, checking galley and page proofs, etc. A period of 1.5 years for this process is not unusual. In my case, the total time elapsed from beginning to end was 3.5 years; I was able to devote but little time to writing during the academic semesters of 1959 and the spring of 1960.

Ernest L. Eliel at the Notre Dame Post Office in 1960, possibly carry-ing part of the Stereochemistry *manuscript.*

The book was an instant success. The world of chemistry, or at least the world of organic chemistry, was hungry for a systematic treatment of stereochemistry. The topic was clearly important in the field of reaction mechanisms, which flourished in the 1950s and 1960s, and there had been no systematic treatment of stereochemistry for two or three decades. The book has gone through 12 printings (it is now out of print), and through 1987 (i.e., in 25 years), over 41,000 copies of the English-language original have been sold. Of the copies sold, about half were hard-bound copies and the other half were various soft-cover international student editions.

The book was translated into Japanese, German, Czech, and Russian. The German translation was undertaken by Arthur Lüttringhaus and Rudolf Kruse, who were kind enough to send me copies of the translated manuscript for comments. It became clear to me that the German language is more precise than English; the meaning of certain slightly ambiguous statements—which I could get away with in the English original—had to be clarified in the German translation. Fortunately, Lüttringhaus is himself an expert in the field of stereochemistry, and his interpretations of the text were almost invariably to the point.

The Czech translation was arranged by my friend Jiri Sicher, mostly because he wanted to avoid seeing the original book pirated in Czechoslovakia. In those days, many of the iron-curtain countries did not adhere to the International Copyright Convention—certainly the Soviet Union did not, as evidenced by the appearance of an unauthorized Russian translation.

The head of the Moscow translation bureau was O. A. Reutov. I first met him at a meeting of the International Union of Pure and Applied Chemistry (IUPAC) in Montreal in 1961, and I told him about the impending appearance of my stereochemistry book. His reaction was: "We shall look at it, and if we like it, we shall translate it." I saw Reutov again at a party at the IUPAC meeting in London in 1963 and asked him if he had translated the book. He replied, "Don't you know. . ." At that point, we became separated somehow (the room was crowded), and I could never locate him again to hear the end of the story.

In fact, I did not know that the book had been translated until 1965, when at the first (officially "zeroth") "Bürgenstock Conference" on stereochemistry I met a Bulgarian chemist who showed me an ugly-looking black-bound book printed on yellowish paper, which, after transliterating the Cyrillic title page, I recognized as a translation of my book. I was subsequently sent a copy and—several years later, when I broached the subject—was offered a royalty of 900 rubles to be paid whenever I cared to claim it in the Soviet Union. The occasion has not arisen yet.

The book vies with my original work on the conformational analysis of mobile systems (*vide supra*) as being my most important contribution to chemistry. I have met hundreds of chemists all over the world who have told me that they learned stereochemistry from my book and that they appreciated the way the subject was presented. Certainly, stereochemistry has flourished during the past 25 years and is now more important than ever, in view of the current wide interest in the synthesis of enantiomerically pure natural products and synthetic pharmaceuticals.

Professorship at Last

Soon after my return to Notre Dame, I learned from Francis X. ("Tim") Bradley, who had become research administrator, that the U.S. Army had available some substantial sums of money for research instruments. He urged me to put in a proposal for a Varian HR-60 NMR instrument, which was funded. Thus, beginning in 1960, we had the opportunity to record proton NMR spectra, which were extremely useful in our research in conformational analysis.

Otherwise the situation at Notre Dame was somewhat bleak. I felt that I should by now have been promoted to professor. My research was going well and was, by 1959, funded by the NSF and the

Ernest L. Eliel, probably in 1960, in front of the Chemistry Hall at Notre Dame, where he worked from 1948 until 1972.

PRF, as well as the AOOR. I had 47 independent publications to my name (nine of which were published in 1958), including several quite important ones. Moreover the NSF fellowship awarded to me involved a measure of recognition. However, my relations with G. Frank D'Alelio, who (against the opposition of most of the chemistry faculty) had become head of the department in 1955 after C. C. Price's departure, were cool. He did not understand my research, and I did not understand his work (which was quite applied). Evidently, he failed to advance my promotion in 1958 and again in 1959.

The situation in the department changed dramatically in 1960, when Frederick D. Rossini arrived at Notre Dame as dean of the College of Science and acting head of the chemistry department. I was one of four faculty members promoted at Notre Dame that summer and felt very proud about it. Whereas up to that point I had considered myself just as an employee of the university and its chemistry department, from then on I felt a personal responsibility about their welfare.

Community Involvement

The early 1960s saw the completion of my stereochemistry book and were a period of consolidation. The hydride work (Schemes 9, 10, and 13) was coming to fruition,[35] and the work on conformational analysis was going in full force. We applied both the NMR and equilibrium

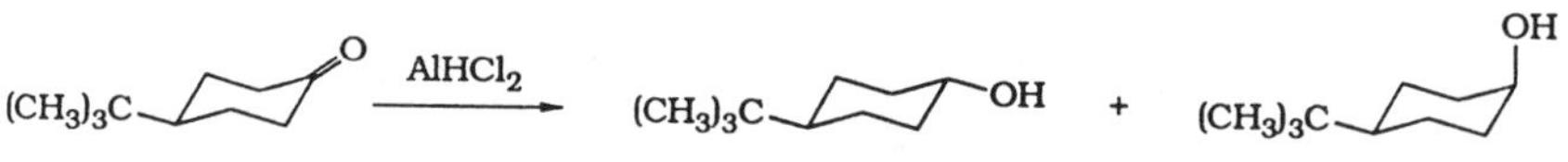

kinetic control:	80%	20%
thermodynamic control: (with excess ketone)	~100%	~0%

Scheme 13

methods (as well as the kinetic method, although doubts about its validity began to surface at this time) to measurements of the conformational free energy preference $-\Delta G°$ (Winstein's A value[29]) of a number of substituents and combinations of substituents. My first review on this subject appeared in 1960[41] and was followed in later years by a number of others. In view of the high "noise level" in the scientific enterprise, I have always felt that reviews are essential in bringing one's work to the lasting attention of the scientific community.

Rupert Ingraham, the McGraw–Hill editor who had been in charge of my stereochemistry book, had warned me that the appearance of the book would change my life. "People will bang on your door and not leave you in peace." He was right. Until 1962 or 1963 (i.e., during the first 15 years of my career), I had had much time to think and read. After my book was published, increased outside demands—to speak, travel, write reviews or book chapters, participate in various organizations, referee manuscripts, review proposals, write recommendations, and, soon, be involved in departmental organization—left me less time for scholarly endeavors. Perhaps this course of events is normal, or perhaps I have sometimes given in too easily to these demands.

In any case, my life during the last 25 or so years of my career has been very different from what it was during the first 15 years. For example, prior to 1962, I was abroad only once (at the IUPAC meeting in Zurich in 1955), but since 1962, I have averaged probably at least one trip abroad per year. When I became a chemist, I did not dream that my profession would afford me an opportunity to travel all over the world and that my travels would usually be enhanced by pleasant personal attention from my colleagues. Whereas others traveling abroad may feel

At Bürgenstock in 1965, Ernest L. Eliel is on the right next to Jack Dunitz; Derek Barton is behind and slightly hidden by Jack.

as outsiders in foreign lands, I have almost always felt right at home everywhere, as a member of the international confraternity of chemists!

One of the public-service jobs I accepted in 1960 was the chairmanship of the St. Joseph Valley Section of the ACS. I had been secretary of the section in 1954–1955 and alternate councilor in 1957–1959. Evidently, I had been able to convince my colleagues that I was a good organizer. The section chairmanship turned out to be just a harbinger of much future service to the ACS, which I have served in various capacities: councilor (1966–1973 and 1975 until the present), member of the executive committee (1966–1968) and chairman (1973–1974) of the Division of Organic Chemistry, and member (1985 until present) and chairman (1987–1989) of the board of directors.

Originally, my interest in the national society was mainly in the area of publications (I have served on the Society Committee on Publications for over 20 years, with some short interruptions, and served as its chairman for 4 years). But through my membership of the ACS council, I began to realize the many important services—to the members, to the profession, and to the science—that the Society pro-

In Heidelberg in 1967 at the time of a professorial offer. Left to right: daughter Carol, Ernest L. Eliel, daughter Ruth, Else and Georg Wittig, and Christel Schenck.

vides. Thus I have been willing to devote increasing amounts of time and effort to it.

Offers to relocate began to arrive in 1962, even before my book had appeared in print. The first offer came from the University of Kansas. I turned down several formal and even more informal offers of this type, because I believed that the purported advantages would be outweighed by the problems of moving. Of course, 10 years later, I did relocate to the University of North Carolina.

Administrative Experience

In 1963, it was clear that F. D. Rossini could not reasonably continue to be both dean of the college and acting head of the chemistry department. There had been no search for an outside head; evidently, Rossini intended to get the department back on an even keel before turning it over to someone else. Among other things, he did something about the abysmal salary levels. In 1959–1960, my annual salary had been $9000; after the offer from the University of Kansas in 1962–1963, it became $14,000.

Early in 1963, Rossini asked me to become department head. I agreed somewhat reluctantly, with the proviso that the start of my term was to be delayed to January 1, 1964, so that I could complete my share of the writing of *Conformational Analysis*,[32] which Norman L. Allinger (then of Wayne State University), Stephen Angyal (of the University of New South Wales), George Morrison (of Leeds University, who had been suggested as a coauthor by D. H. R. Barton), and I had launched in 1961, even before my stereochemistry book had appeared.

The 3 years (1964–1966) of my department headship were a good educational experience for me, even though the autonomy of heads at Notre Dame at that time was quite limited, especially in budgetary matters. I decided to add a biochemical component (which was missing up to that time at Notre Dame) to the chemistry department and succeeded in hiring two good young biochemists who were willing to start from scratch. I renovated a laboratory for them, and I arranged to buy most of the equipment they needed. I was involved also in the first denial of tenure (at least since I arrived in 1948) in the department and managed to attract an excellent inorganic chemist, Tom Fehlner, who later (1982–1988) became chairman of the department. Last but not least, I believe I was influential in securing the award of a science development grant to Notre Dame. (The fact that the president of the university, the Reverend Theodore M. Hesburgh, was a member of the National Science Foundation Board also helped, surely!) This grant led

Ernest L. Eliel in the old laboratory at Notre Dame, around 1966.

to the addition of two more very good biochemists and several other young faculty members in the late 1960s.

However, I failed in one major endeavor, that of constructing a new building. The old building, in which most of our research was done, dated back to the 1920s or 1930s, and an NSF staff member who inspected it in 1964 after minor renovations—mostly of the hoods and benches—declared that it was a miracle that one could do good research in such a building! The new building was planned, but it was not built until 1982. The poor state of the research facilities during the 1970s led to a major decline in the number of graduate students and the department as a whole during that period; this decline has been reversed only recently, largely through Tom Fehlner's efforts, after the new building was finally built.

The science development program of the NSF was designed to "increase the number of truly excellent science departments in the U.S." Unfortunately, the program failed largely. The Notre Dame experience was typical of that of a number of other universities that received science development grants. The fact remains that one can help only those who help themselves. The University of North Carolina, or at least its chemistry department, was one of the few universities where the program produced clear and lasting results.

On the whole, I quite enjoyed the 3 years when I was department head. I realized that I lacked some of the talents—notably those of adroitness and persuasiveness with superiors—needed by a successful university administrator, and I have resisted, I think wisely, several tentative offers to become a department chairman or dean elsewhere since then.

My research suffered somewhat, though not greatly, during the 3 years when I had to devote much time to departmental administration. The work on the free-radical problem was terminated in 1965—perhaps for the better. The work on hydride and conformational problems continued unabated, thanks in part to several excellent predoctoral and postdoctoral collaborators.

In fact, several new projects were broached during that period. Siegfried Schroeter (who, until his untimely death in 1988, was a manager at the General Electric Company) used Raney nickel

Ernest L. Eliel with Ephraim Katchalski at Notre Dame in 1966.

oxidation–reductions to equilibrate substituted cyclohexanols and thereby to determine the conformational energy, $\Delta G^o_{a \rightleftharpoons e}$ of the OH group in a variety of solvents. (The a and e in the subscript stand for axial and equatorial, respectively.) Eugene Gilbert later used this method for a careful determination of $\Delta H^o_{a \rightleftharpoons e}$ and $\Delta S^o_{a \rightleftharpoons e}$ of the same group (Scheme 14 and Table I). Hydrogen bonding, especially to the hydroxyl oxygen, significantly enhances $-\Delta G^o_{OH}$.[42]

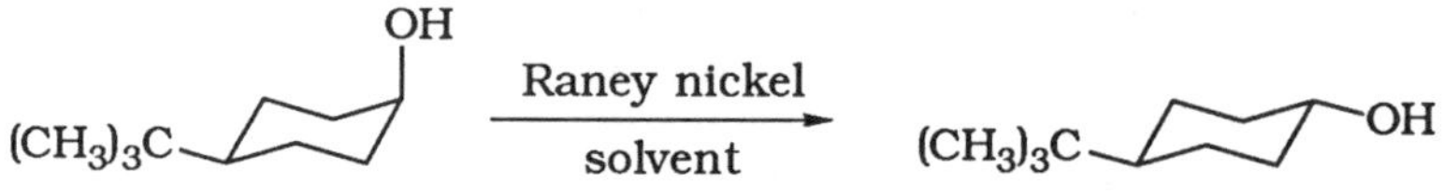

Scheme 14

Table I. Effect of Solvents on Conformational Energy of OH Group

Solvent	$-\Delta G^o$	$-\Delta H^o$	ΔS^o
Cyclohexane			
0.05 M	0.60	0.58	0.06
0.10 M	0.61	0.61	−0.02
0.20 M	0.61	0.63	−0.06
1,2-Dimethoxyethane	0.74	0.83	−0.30
2-Propanol	0.95	1.09	−0.46
2-Methyl-2-propanol	0.95	1.18	−0.76

NOTE: Values of $-\Delta G^o$ and $-\Delta H^o$ are expressed in kilocalories per mole, and values of ΔS^o are expressed in calories per mole per kelvin.

T. Brett studied similar equilibrations by using $RCHOAlCl_2$ complexes in which the $OAlCl_2$ substituent is almost entirely on the equatorial side (Scheme 15).[43] Earlier we had found a way to reduce monothioketals to hydroxythioethers (Scheme 8), and we now discovered that their reduction with calcium in liquid ammonia would lead cleanly to alkoxymercaptans (and in the case of dithioketals, to alkylthiomercaptans).[44]

The year 1967 also saw the appearance of Volumes 1 and 2 of *Topics in Stereochemistry*. The publication of this series, which by now comprises 19 volumes, came as an aftermath of the publication of *Conformational Analysis*, which brought me in close scientific contact with Lou Allinger, who was then at Wayne State University. Allinger was coeditor of the series through Volume 16. (Samuel H. Wilen of the City University of New York, who was a postdoctoral collaborator in my

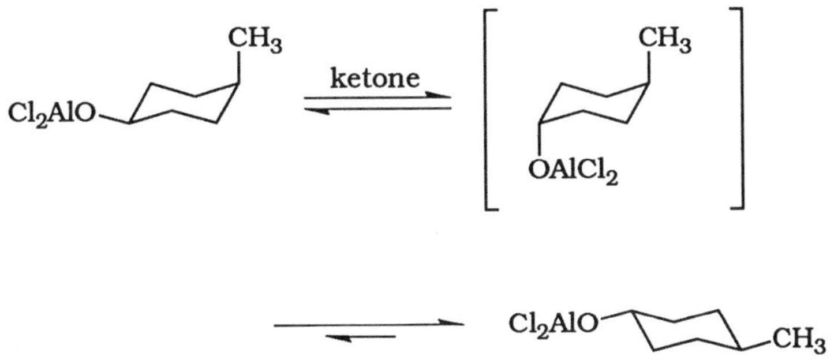

Scheme 15

laboratory in 1955–1957, is now coeditor.) The series has been well received, and the first two volumes sold almost 2000 copies each. In recent years, unfortunately, a vicious spiral in price (up) and the number of copies sold (down) has set in, with the price having increased from $12.50 (Volume 1) to $100 (Volume 19) and sales now down to less than 1000 copies. One wishes book publishers would find a way to break out of this spiral!

Conformational Analysis of Saturated Heterocyclic Compounds: Part I

Undoubtedly, our most important effort during that period involved the extension of conformational analysis to saturated heterocyclic compounds. The idea of investigating this area had been planted in my mind through conversations with S. J. Angyal, with whom I had a number of contacts during the early 1960s through our coauthorship of *Conformational Analysis* and who, as a carbohydrate chemist, was obviously interested in the area. The pioneering work was done by Sister Margaret Knoeber with an investigation of 1,3-dioxanes. Diastereomers of 1,3-dioxanes can be equilibrated easily by treatment with anhydrous acid (Scheme 16).[45]

For this equilibration, we first used boron trifluoride etherate as the catalyst, but as a result of my attending the 1965 Gordon Conference on Hydrocarbon Chemistry, I learned about the use of beaded poly(styrene sulfonic acid) (Amberlyst-15) as an effective heterogeneous acidic catalyst. We soon switched to this acid because of its easy removal from reaction solutions.

The dioxane work brought two surprises. One was that the equilibrium constants for a variety of 2-alkyl-5-*tert*-butyl-substituted dioxanes (Scheme 16; 6 ⇆ 7) were about the same, $-\Delta G^\circ = 1.4 \pm 0.1$

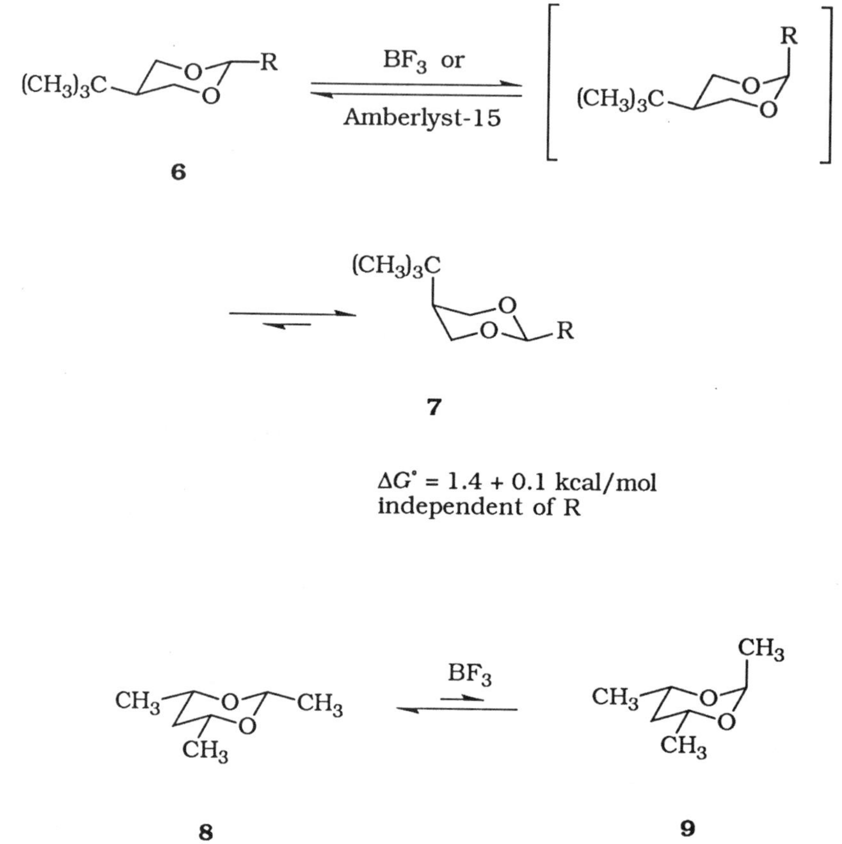

6

7

$$\Delta G^\bullet = 1.4 + 0.1 \text{ kcal/mol}$$
independent of R

8   **9**

$$\Delta G^\bullet = 4.0 \text{ kcal/mol}$$

Scheme 16

kcal/mol, regardless of the nature of the alkyl group at C-2. We soon found out by proton NMR spectroscopy (about this time we had acquired a Varian A-60A instrument) that in all cases the *cis* isomer existed with the alkyl group at C-2 in the equatorial position and the *tert*-butyl group at C-5 (which we had hoped would serve as a conformation-holding substituent) in the axial position. Evidently, in all cases we observed an equilibrium between an equatorial (*trans*) and axial (*cis*) *tert*-butyl group at C-5!

Surprisingly, thus the interaction of the axial *tert*-butyl group with the oxygens at positions 1 and 3 (including their lone pairs) was much smaller than the corresponding interaction with the C–H(ax) in axial *tert*-butylcyclohexane ($-\Delta G^\circ = 4.9$ kcal/mol[46]). This result demonstrated that the steric requirement of a lone pair was quite small, a fact that had not been clear at all in the mid-1960s. In contrast to alkyl groups at C-5, alkyl groups at C-2 in 1,3-dioxanes have very large $-\Delta G°$ values (e.g., 4.0 kcal/mol for CH_3 versus 1.74 kcal/mol for cyclohexane), presumably because the interactions with the axial hydrogens at C-4 and C-6 are very severe. We ascribed this severe interaction to the shortness of the C—O—C bond distances, along with the observed puckering of the 1,3-dioxane ring at C-2,[47] which makes an axial substituent at this position lean inward. The primitive notion held until the mid-1960s[32] that stability in substituted saturated heterocyclic compounds would mimic that in analogous isosteric carbocyclic compounds is evidently incorrect!

Establishment of the position of the equilibrium shown at the bottom of Scheme 16 was a very difficult task. The analysis of the equilibrium mixtures was carried out by gas chromatography (we had acquired gas chromatographs in the late 1950s), but of course, the chromatograms had to be calibrated with authentic samples. Unfortunately, however, the condensation of *cis*-2,4-pentanediol ($CH_3CHOHCH_2CHOHCH_3$) with acetaldehyde to give compounds 8 and 9 (the *cis*-4,6-dimethyl substituents served as holding groups because a *tert*-butyl group at C-5 evidently would not do) requires an acid catalyst and therefore proceeds under equilibrium conditions in which isomer 9 is present in negligible amounts only.

Fortunately, at about this time, the late Jiri Sicher visited from Czechoslovakia. Sicher was apprised of the problem and suggested that we use a very short reaction time in the preparation of 8 and 9, even if that led to incomplete conversion of the starting glycol and aldehyde. He opined that shortening the reaction time would lead to at least partial kinetic control and that the thermodynamically less stable product might (as so often it is) be favored kinetically. The idea worked, though only after a fashion: The product obtained on incomplete reaction contained about 1–2% of 9, and Sister Margaret Knoeber (who, before the days of ecumenism, was known by her religious name, Sister Carmeline) spent many hours in front of a preparative chromatograph to isolate enough of 9 to characterize it by its NMR spectrum and chromatographic retention time. After we published this work, we learned that similar results with respect to the 6 ⇆ 7 equilibrium had been obtained by Frank Riddell and Michael Robinson in Oxford.[48]

A Sabbatical Year at the ETH in Zurich

In 1967 I received a second NSF senior postdoctoral fellowship and a leave of absence from the university for the period 1967–1968. Eva and I decided to spend the year with our two teen-aged daughters in

Ernest L. Eliel on the S.S. Constitution *on the way to Zurich in the summer of 1967 with his daughters Ruth (left) and Carol.*

Zurich, at the Eidgenösische Technische Hochschule (ETH). I thus had the delightful experience of spending a year with our old friend Vlado Prelog, participating in Prelog's research seminars, receiving many private tutorials (for which he almost always had time, despite his many activities), and listening to his inexhaustible repertoire of entertaining stories that often made telling philosophical or scientific points. Moreover, the year at ETH gave me the chance to get to know well the other members of the "Zurich Mafia", notably Duilio Arigoni, Jack Dunitz, Albert Eschenmoser, and André Dreiding (at the University of Zurich).

My association with Duilio Arigoni was particularly fruitful. I had gone to Zurich purposely to study the phenomenon of heterotopism (also called prochirality,[49] perhaps infelicitously), which had just then been put on the map, so to speak, by a pioneering article by Mislow and Raban[50] that we had the good sense to publish as Chapter 1 in Volume 1 of *Topics in Stereochemistry*. Because of the important applications of heterotopism in the elucidation of biochemical reaction paths, the logical place to pursue such a study would have been in the biochemical area. However, I did not have a burning desire to get into the necessary experimental work. In fact, I probably missed my last chance, during this sabbatical, to learn to do something radically different from what I had done earlier. The next best thing was to write a review on the subject. A review seemed particularly appropriate,

Ernest L. Eliel with J. Dunitz (center) and D. Arigoni (right) at Vlado Prelog's 80th birthday party in Zurich in 1986.

because in fact, most of the studies of the "cryptochirality" of "prochiral" CH_2 groups (i.e., groups in which the two hydrogens are heterotopic) involved CHD analogs. Thus, there was a direct connection to my first "love", the chirality of $C_6H_5CHDCH_3$, which, as I mentioned earlier, I had not pursued during the intervening 20 or so years.

I managed to interest Arigoni in coauthoring a chapter[11] on the chirality of RCHDR' compounds for *Topics in Stereochemistry*. Thus I was introduced to a fascinating set of experiments, including J. W. Cornforth's beautiful elucidation of the stereochemical course of the synthesis of squalene from mevalonic acid.

Collaborating with Arigoni was an interesting experience. He had a complete command of most of the literature pertinent to our chapter. Thus (except for the early parts of the chapter that related to the subject of my 1949 work) we would usually start by my receiving a set of references, which I would then proceed to read and subsequently discuss with Duilio. After our discussion, I was ready to write the appropriate section. Arigoni would then read the written material and criticize it incisively. His criticisms led to rewriting (by me!) and much improvement.

Arigoni himself never wrote anything; he claimed that he lacked facility with written English, although his spoken English is excellent. I cannot corroborate his claim, however, because during our critique ses-

Ernest L. Eliel at the Nomenclature Workshop at the Ciba Foundation in London in 1968. Top, with Kurt Mislow (left). Bottom, with R. S. Cahn (left).

sions, he not only corrected factual or interpretational errors but also often criticized details of language and took exception (almost always with good reason) to certain words and expressions, replacing them with more appropriate ones. There is no question that he has an excellent feel for the English language; nonetheless, he has an established reputation for not writing up his own excellent research for publication.

In addition to writing a chapter, taking several lecture trips to various parts of Europe, and attending the famous Bürgenstock conference on stereochemistry (so called because it annually takes place at a very posh group of hotels on a rock called the Bürgenstock high above Lake Lucerne), I also wrote several other papers. By 1967, we had learned that neither the kinetic method of conformational analysis[51] nor the NMR spectroscopic method using model compounds[38,39,52,53] was very reliable. Therefore, we had concentrated broadly on the equilibrium method,[54,55] as, for example, with the 1,3-dioxanes (Scheme 16).[45]

One of the equilibria we had studied was that of 2-alkoxy- and 2-alkylthiooxans,[55] compounds that manifest the anomeric effect seen in analogous sugars (i.e., a preference for axial over equatorial RO and RS groups). This problem had been discussed by Lemieux[56] and later by Havinga and his group,[57] but we felt we could throw new light on it. By contemplating models, we realized that whenever there were *syn*-parallel electron pairs (Scheme 17), destabilization would result. In fact, the crystallographic literature suggested that in β-glycosides, the aglycone is almost always so oriented as to make the number of such

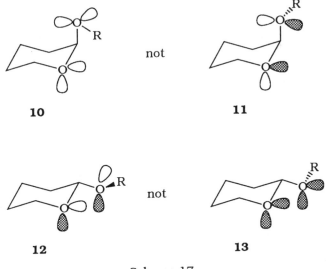

Scheme 17

interactions zero (**10**) rather than one (**11**), whereas in the α-anomer, the orientation is such as to generate one (**12**) rather than two (**13**) such interactions. We then looked for similar interactions (or rather their avoidance) in other compounds and soon found them in N,N'-dialkyl-1,3-diazanes (Scheme 18), in which one of the N-alkyl groups is axial[58] (although not as predominantly as we had thought originally[59]).

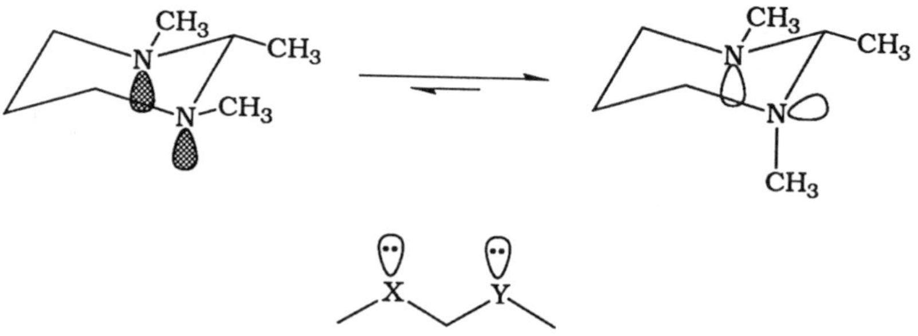

Scheme 18

I discussed this topic with a colleague during a visit to the Dundee campus of St. Andrews University (the home of Douglas Neilson, who had worked in my laboratory the year before[53]). We left a diagram (Scheme 18, bottom) on the blackboard, and the next morning, I saw that someone had written "Brer Rabbit" underneath the picture. From that moment, I dubbed the phenomenon the "rabbit-ear effect".[58–61] In fact, soon thereafter, as a result of a lecture I gave at the Centennial of the Chemical Society of Lund (Sweden), I wrote a review on this effect[60] and pointed out that this phenomenon had also been seen in acyclic compounds, such as $CH_3OCH_2OCH_3$ and the corresponding polymer poly(oxymethylene) $[-(CH_2O)-]_n$, which, unlike linear polyethylene, has a helical conformation because of the rabbit-ear effect. The somewhat mischievous editor of *Kemisk Tidskrift*, in which the review appeared, put a full-size picture of a rabbit (with prominent ears!) on the cover page of the issue in which my article appeared (*see* photo).[60]

I should have known better. In fact I should have heeded the warning of Jack Roberts, who (openly) refereed the communication about the effect in the *Journal of the American Chemical Society*.[58] Roberts felt that the terminology was inappropriate, especially in view of the uncertain etiology of the phenomenon. Indeed, although I had made no such claims, it was soon assumed that I implied the effect to be of quantum mechanical origin. The quantum mechanics specialists delighted in showing that it was not so but that, in fact, the effect was due (at least in part) to a favorable p–σ^* bond overlap (or double-bond–no-bond resonance) (Scheme 19).[62]

The cover of Kemisk Tidskrift *No. 6–7, 1969.*

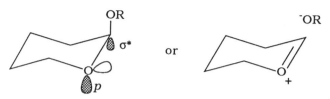

Scheme 19

However, part of the effect is due also to dipole–dipole interaction, and parallel electron pairs, as shown in Scheme 18, imply repulsive dipoles. To the best of my knowledge, the crucial experiment—in which the electron pairs are parallel but the dipoles are attractive—has never been performed. In any case, the term rabbit-ear effect has now been struck from the literature, and by arrangement with Ray Lemieux, we have gone back to calling the phenomenon the generalized anomeric effect.[61]

One of the unexpectedly exciting trips I took during my sojourn in Zurich was to the annual meeting of the French Chemical Society in Montpellier in the spring of 1968. I had been asked to deliver a plenary lecture about our work on conformational analysis in the excellent company of R. B. Woodward, J. Wilkinson, and E. Katchalski, among others. During the flight to the meeting, I was told that an airline strike was in the offing and that my return might be impeded. Indeed, on the Thursday of my stay in Montpellier (the meeting was scheduled to end at noon on Saturday), the airlines went on strike, and Mme. Mousseron, one of the local organizers of the Congress, immediately exchanged our airline tickets for train tickets. Unfortunately, the next day (Friday), a general strike was declared and the trains stopped, too. (These stoppages occurred while the students in Paris were heaving cobblestones at the police, and everyone was very nervous: If the workers had joined the students in a common cause, and this seemed possible, revolution would probably have ensued.) After consulting with Bob Woodward, I called Mme. Mousseron (although Woodward's French was better than mine, he refused to speak the language), and she promised us a rental car the next morning. On Woodward's advice, I asked her for a Peugeot 404 (which accommodated five people), because Woodward contemplated that, during the drive to Geneva, Switzerland, we should take along Wilkinson and his wife and Katchalski.

Katchalski had just arrived—he was delayed by the strike—and was to give his lecture on the Saturday morning. He did not seem concerned about the situation at all (obviously, he had seen worse during the wars in Israel) and was not even sure if he wanted to come with us.

Early the next morning, the car was delivered, but it was a Renault, not a Peugeot, and had room for only four people. Woodward

Return from Montpellier in 1968. Left to right: R. B. Woodward, Mrs. Wilkinson, E. Katchalski, and G. Wilkinson.

asked (or perhaps I should say, ordered) me to send it back and to insist on the Peugeot. After some remonstration, I complied. While Woodward, Katchalski, and the Wilkinsons left for the university, I anxiously waited for the Peugeot, which finally arrived at midmorning. Expecting my four passengers to be ready, I quickly drove to the university, only to discover, to my dismay, that Katchalski was still lecturing, already more than 15 minutes in overtime and oblivious to the general situation and to the urge everyone (including the French participants) felt to get away. After he stopped, there was a further delay when Woodward gave out awards to the best young French chemists. Fortunately, the president of the Societe Chimique de France himself was anxious to end the proceedings, and so we left before noon. (I had made sure that we had food in the car, as well as a full tank of gasoline.)

To my surprise, Woodward insisted on driving, even though he was clearly not used to a stick shift any more, and he stayed in the driver's seat for the entire 8-hour trip to Geneva. I realized later that he would not entrust his life to anyone else! I took over the wheel only after he got off at the Geneva railroad station to take the train to his laboratory in Basel (he did not like flying). The rest of us went on our respective ways from the airport, where we turned in the car.

An earlier trip to France had been to the Collège de France in Paris at the invitation of Alain Horeau. There (as later in Montpellier) I lectured on conformational analysis in French (which I remember from my high-school days and from French conversation lessons I had as a teenager). Although I tend to be self-critical, I felt that the lecture had gone well, both content- and language-wise, and so I was pleased when I heard Bianca Tchoubar (who had come from the Natural Products Laboratory in Gif-sur-Yvette to hear my lecture) remark to Horeau that my French had been quite good and that, when I did not know a word, I found a way around it. Unfortunately, my satisfaction was deflated when Horeau himself (in typical Parisian fashion) paid me the following backhanded compliment: "Your French was quite good and would be even better if you occasionally read a French novel." I wished I had time to read French novels!

Conformational Analysis of Saturated Heterocyclic Compounds: Part II

As a result of our work on the conformational analysis of saturated heterocyclic compounds,[45,55,63–66] which was carried out with a number of excellent collaborators, including Ed Willy, Robert O. (Bob) Hutchins (later chairman of the Chemistry Department at Drexel University), Moses Kaloustian (now at Fordham University), O. Hofer (now at the University of Vienna), William F. (Bill) Bailey (now at the University of Connecticut), and Slayton A. Evans (now at the University of North Carolina), among others, as well as, in some instances, my brilliant colleague Gerhard Binsch (now at the University of Munich), we discovered several interesting reactions of these compounds.

In 1970, Franz Nader, a postdoctoral fellow from the University of Heidelberg, studied the reaction of conformationally well-defined ortho esters with Grignard reagents (Scheme 20). He found that axially substituted ortho esters (14) reacted smoothly to give mostly axially substituted 2-alkyl-1,3-dioxanes (15) (the same compounds that Sister Margaret Knoeber had found so elusive), whereas equatorially substituted ortho esters (16) reacted sluggishly and possibly only after prior epimerization.[67] This difference in reactivity was recognized quickly by my co-workers (who learned of the result before I did; perhaps I was away on one of my many trips!) as a manifestation of the principle of least motion: departure of the axial (but not equatorial) alkoxy group after coordination to $RMgX$ or MgX_2 (from the Schlenk equilibrium) leads smoothly to intermediate 17, in which the empty orbital of the carbocation overlaps with one pair of p electrons of each oxygen. Top-side

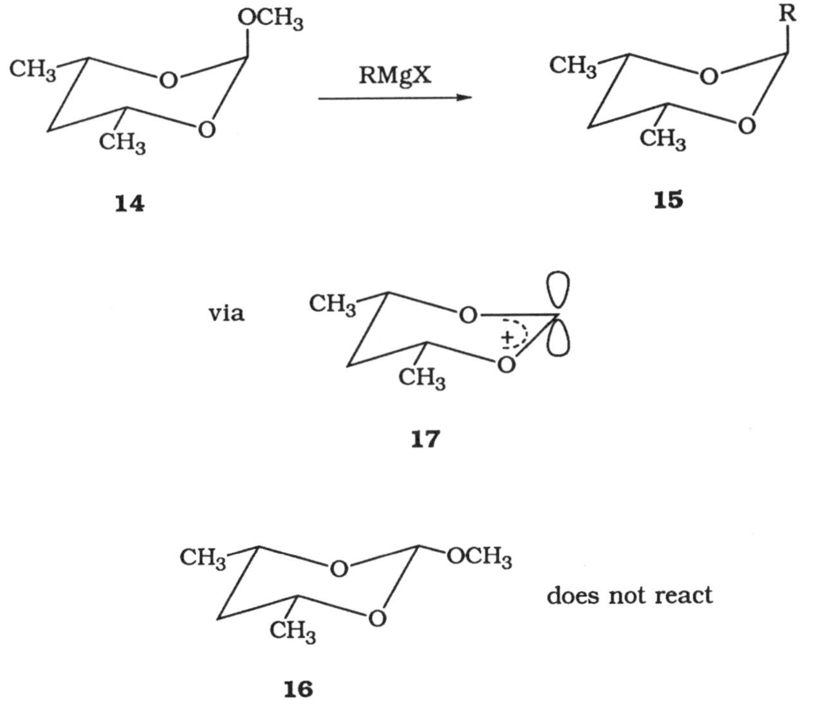

Scheme 20

approach of the alkyl moiety of the Grignard reagent then leads to **15** (*trans-chair* form), whereas bottom-side approach would lead to the boat form of the *cis* epimer through a sterically very unfavorable transition state. It is, of course, fortunate that the method leads to the otherwise inaccessible *trans* isomer; the *cis* isomer (equatorial R group) is readily available by an aldehyde–diol condensation, as explained earlier.

No sooner had the preliminary account of these findings been published than an even more important discovery was made by Armando Hartmann, one of the members of the already mentioned group of my very excellent collaborators during the late 1960s. I had met Armando when I gave a 6-week course on stereochemistry at San Marcos University in Lima, Peru, during the summer of 1966. (As a result of having been an undergraduate at the University of Havana, I was easily able to lecture in Spanish and was called upon to give courses in the Spanish-speaking world from time to time. Most of the other stands were in Mexico; in recent years, I have also given courses in Spain and have made several lecture trips to that country.) The course in Lima had not been very successful on the whole, because most of the participants were poorly prepared, but Hartmann stood out both in performance and in motivation, and I persuaded him to pursue graduate studies at the University of Notre Dame the following year (1967).

Hartmann was following up Hutchins' work on conformational equilibria in heterosubstituted 1,3-dithianes (the equilibrations were not very clean, and the work has never been published) when one day he told me that he wanted to study the Corey–Seebach reaction in conformationally locked 1,3-dithianes, such as the cis-4,6-dimethyl derivative (Scheme 21). I was not very enthusiastic about the plan initially, and to this day I do not know what motivated Hartmann's suggestion: whether he saw some analogy with the then-just-emerging work of Fraser and Durst[68] on the selective abstraction of diastereotopic hydrogen atoms in cyclic dibenzyl sulfoxides, whether he had some prescience of the underlying theoretical arguments, or whether (as I suspected) he simply wanted to do "his own thing". However, I have a principle that, in order to encourage originality among my co-workers (and also to keep up their morale!), if they make a suggestion that is neither clearly wrong nor obviously very expensive in terms of time or materials, I allow them to follow up that suggestion.

In Hartmann's case, the results proved startling. When cis-4,6-dimethyl-1,3-dithiane was treated with butyllithium and then with D_2O, the only product to be seen (by proton NMR spectroscopy) was the one with equatorial deuterium (Scheme 21). Through control experiments, we estimated the stereoselectivity of the reaction to be at least 95%.[69] Subsequently, the lithium derivative was treated with methyl iodide; in this case (Scheme 21), the product (18) was over 99% diastereomerically pure, as shown by gas chromatography. When this equatorial (all-cis) isomer (18) was again converted to the lithium derivative with butyllithium and then quenched with water, it was integrally converted to the axial isomer 19. Presumably, in both cases the carbanion–lithium pair was exclusively equatorial, and the subsequent electrophilic substitution by D or CH_3 proceeded with retention of configuration. Thus, the two stereoisomers, 18 and 19, can be produced at will, both in complete stereochemical purity.

We thought at first that the large equatorial preference of the carbanion pair was a result of ion pairing with Li^+ on the more accessible equatorial side, but later, spectroscopic experiments convinced us that this was not so (because solvent-separated ion pairs behave in the same way as tight ion pairs) but that the equatorial preference was intrinsic to the carbanion itself.[70] In the meantime, it had been pointed out[71] that the equatorial preference of the 1,3-dithianyl-2-carbanion and the previously discussed axial preference of the 1,3-dioxanyl-2-carbocation are opposite manifestations of the same phenomenon, namely overlap of the axial orbital at C-2 with the antiperiplanar unshared electrons on oxygen or sulfur (Scheme 22). This overlap is

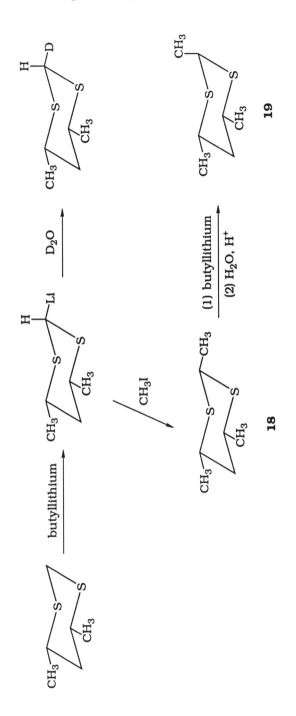

Scheme 21

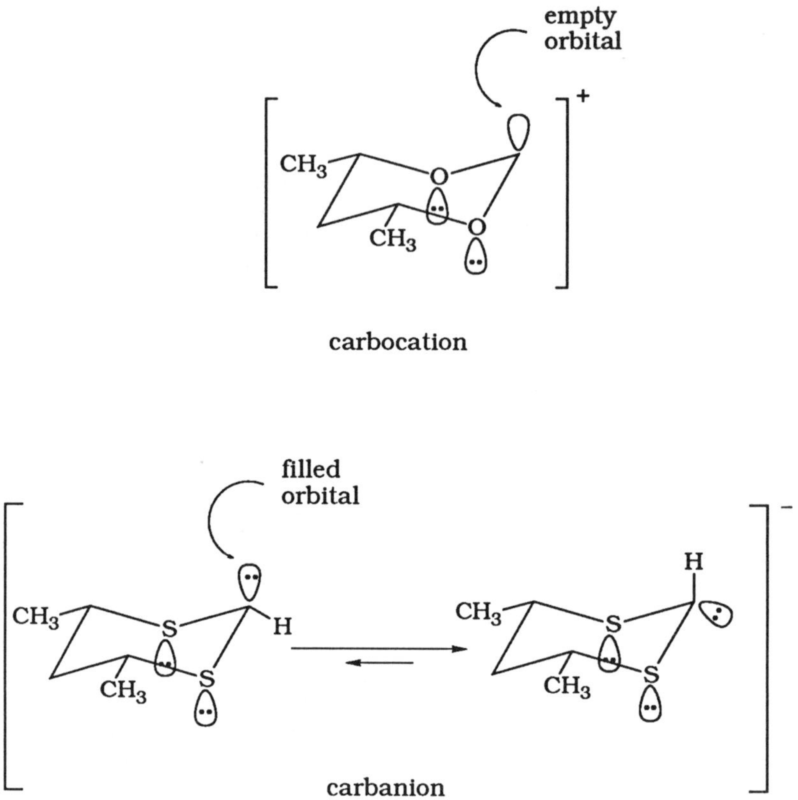

carbocation

carbanion

Scheme 22

favorable for the unfilled orbital in the carbocation (which therefore prefers the axial orientation) but unfavorable for the filled orbital of the carbanion (which thus avoids the axial position). Lehn and Wipff[72] calculated the preference for the equatorial carbanion over the axial carbanion to be about 10 kcal/mol, a figure that explains handsomely why we were unable to observe any products formed from the axial species. Later, elegant crystallographic work[73] confirmed the equatorial nature of the lithium in 2-phenyl-1,3-dithianyllithium. Incidentally, this confirmation also supported the conclusion that ion-pair formation was not primarily responsible for this observation.

We realized almost immediately that the very high equatorial preference of the anion pair (leading to an enzymelike stereoselectivity in the processes shown in Scheme 21) might be harnessed to a highly stereoselective (asymmetric) synthesis.[74] However, it took us nearly 5 years to achieve the goal of developing conformational principles into a useful synthetic application. Meanwhile other interesting things were

The organizers at the Table Ronde Roussel-Uclaf in 1970: Ernest L. Eliel, J.-M. Lehn (center), and J. Dunitz (right).

happening in the laboratory; I shall turn to these before coming back to the sequel of the lithiodithiane work, but first I must mention my change of location.

New Walls

As I already mentioned, during the late 1960s, I had received several attractive invitations to look into the possibility of relocating. These invitations included several actual offers of positions, one of them from the University of North Carolina. However, in part because I was then quite happy at Notre Dame and in part because I felt responsible for the successful implementation of the Science Development grant, I turned down all such offers.

By the early 1970s, the situation had changed. The heady period of science during the 1960s was evidently coming to an end, as was Father Hesburgh's term on the National Science Board. It became painfully obvious that the promised new chemistry building at Notre Dame

"Goodbye Notre Dame", 1972. Left to right: Ernest L. Eliel, Gerhard Binsch, and John Magee.

would not be built, at least not at any time soon. On a different front, the ecumenism that had developed so mightily under Pope John XXIII was beginning to abate, and the University of Notre Dame became preoccupied once again with its Catholic character and that of its faculty. An endowed professorship had been established in the Chemistry Department, but it was not offered to me. In the summer of 1971, in response to a second offer, I accepted a W. R. Kenan, Jr., professorship at the University of North Carolina, with the move to be effective on July 1, 1972. In the interim, in April 1972, I was elected a member of the National Academy of Sciences. Although I was, of course, very happy about this singular honor, my colleagues at Notre Dame, for obvious reasons, viewed it with mixed emotions.

Not wishing to use this account as a means of extolling the virtues of our department at the University of North Carolina, I shall say simply that I have never regretted my move, even for a moment. After 17 years, I feel that the department is still on the way up, thanks to the continuing excellent leadership and a very hard working and scientifically outstanding faculty. The strength of the University of North Carolina is that it encourages departments to look after themselves and rewards those that do.

^{13}C NMR Spectroscopy

Back to chemistry. About 1970, Wesley Bentrude of the University of Utah, a fellow researcher in the area of conformational analysis (principally of phosphorus compounds), drew my attention to some very exciting work on ^{13}C NMR spectroscopy as related to conformation then going on in the laboratory of his colleague David Grant.[75] He suggested that I collaborate with Grant, and indeed, David Grant was quite willing to enter into a collaboration. The first ^{13}C NMR spectra of 1,3-dioxanes were soon published.[76] (Later, while still at Notre Dame, I also collaborated with Ernest Wenkert, who was then at Indiana University, on ^{13}C NMR spectroscopy.)

The results of the ^{13}C NMR spectroscopic work were surprising. Although it was generally accepted that axial methyl groups in cyclohexanes resonate at higher field than equatorial ones,[75] thanks to the "γ-compression effect", the opposite was true for methyl substituents at C-5 in conformationally locked 1,3-dioxanes, such as the *cis*- and *trans*-2-*tert*-butyl-5-methyl-1,3-dioxanes.[76] Even trying to understand this reversal required a lot of additional work, which was done collaboratively by Grant's and Wenkert's groups and our own.

On the basis of the results obtained, David Grant and I tried to formulate a theory for the earlier mentioned reversal of chemical shift of methyl groups at C-5 in 1,3-dioxanes. I shall not repeat the interpretation[77] here, because it is still quite controversial and certainly in part incorrect. However, it led to a highly cited paper,[77] which boasts of no fewer than 12 authors. In my opinion, David Grant provided the major impetus for this paper and his name should have come first, but in his typical modesty, he insisted that mine be put at the beginning because I had originated the problem. I am afraid I gave in to his pressure too easily. Because *Science Citation Index* lists citations of papers only under the first author, I have received credit that should have gone to Grant.

When I moved to North Carolina, my friend Günther O. Schenck (of the Max Planck Institute for Radiation Chemistry, Mülheim) sent me a note, which, after translation from the German, said, "It is healthy to look at new wallpaper from time to time." He was right. The move to North Carolina brought a spurt of new research activity, partly because I needed to prove myself, partly because no one outside the department yet knew me so that I was not burdened with university committee work, partly because the department had given me money for additional postdoctoral collaborators, and partly because the grants I transferred from Notre Dame went further at a state university than they had at a private university (where both tuition costs for graduate students and the overhead rate had been quite high).

Even before I arrived I contributed to a proposal submitted to the NSF for a 100-MHz NMR instrument equipped with Fourier-transform capabilities to record ^{13}C NMR spectra. The grant was awarded, and soon after my arrival, I consulted with the late Charles Reilley and with Maurice (Brook) Brookhart about which instrument to buy. Reilley, an almost universal genius, knew the insides and outsides of all the available instruments intimately and felt strongly that we should buy a Varian XL-100, even though he knew full well that Varian would sell us an incomplete instrument, as they were in the habit of doing in those years. (Indeed, it took almost a year after delivery of the instrument before all the parts were in hand.) The problem was that we did not have quite enough money. My contribution to the purchase came in the final round of negotiations with the Varian technical salesman. When he named what he said was his lowest possible price, I told him that he had just sold us the competition's instrument. He became pale, ran to the telephone to communicate with his office, and soon came back with a price that we could afford.

The instrument lasted for 14 years and provided an absolutely enormous boost to our research. Much of that boost is due to the fact that even before the instrument arrived we engaged an outstanding person, David Harris, to supervise and operate it. Harris has had major input to many of our papers, not only by keeping our departmental NMR instruments in running order almost 100% of the time and either operating them or supervising and training student operators, but also by introducing us to many new and important NMR spectroscopic techniques. All the credit he normally wants and gets is an acknowledgment in our papers.

With the new instrument in place, we found that ^{13}C NMR spectroscopy was an excellent tool to assess conformational equilibria by recording ^{13}C NMR spectra at low temperatures. It had long been known[78] that the position of the equilibrium shown in Scheme 8 can be assessed by decreasing the rate of interconversion of the conformers (by lowering the temperature) to the point at which they can be seen separately in the NMR spectrum (i.e., when their rate of interconversion is low compared with the difference in chemical shifts of salient protons, e.g., CHOH). However, because of the general poor resolution of the proton spectra and the width of the (coupled) signals, this method is not ideal. Moreover, a proton spectrum tends to produce only a single piece of data. In contrast, ^{13}C NMR signals are sharp, and often the signals for several carbon atoms in the two conformers are well resolved, so that several sets of data can be obtained from a single spectrum. Difficulties due to unequal relaxation times and nuclear Overhauser effects can be minimized by observing corresponding carbon atoms in the two conformers. Using this method, we performed conformational analyses of a variety of substituted cyclohexanes,[79] as well as a number of saturated heterocyclic compounds, which will be discussed later.

One of the problems we hoped to solve by ^{13}C NMR spectroscopy was that of the position of the N–CH_3 and N–H axial–equatorial equilibria in piperidines and N-methylpiperidines (Scheme 23; 20 and 21). Whereas with N-methylpiperidine the equilibrium was clearly on the equatorial side, estimates of $-\Delta G°$ varied widely, from -0.65 to -1.61 kcal/mol (corresponding to 75–94% of the equatorial isomer).[80] With piperidine itself, it was not even clear whether the N–H axial or N–H equatorial conformation predominates (although the difference in free energy was evidently small, $\Delta G° \leq 0.6$ kcal/mol[81]); in fact, in 1975, what one might call a "review war" occurred, in which Alan Katritzky[82] argued forcefully on the side of predominantly equatorial N–H and Joseph Lambert[83] nearly equally forcefully argued on the side of axial N–H.

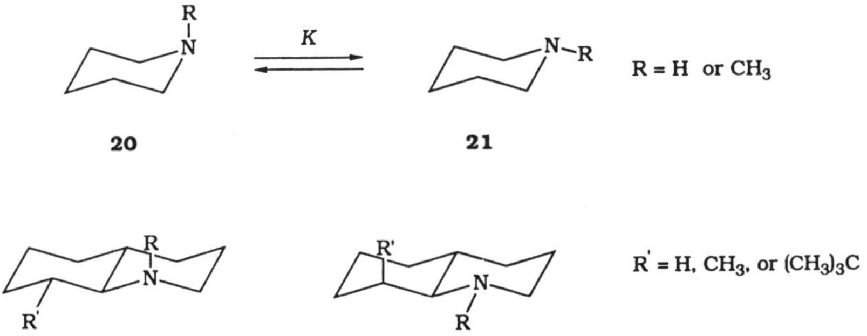

Scheme 23

Because the N–R (R = CH_3 or H) axial–equatorial equilibrium is too fast even on the NMR time scale to be readily observed directly (i.e., by "freezing out" the conformers; however, *see* reference 81) and because other physical methods (such as infrared spectroscopy or measurement of Kerr constants or dipole moments) had proved unreliable, indirect approaches had to be devised. The approach we chose is shown in Scheme 23 (22 and 23). With an appropriately large R' group at position 8 in the carbocyclic ring of a *trans*-decahydroquinoline, the R substituent on nitrogen would be axial when R' is equatorial (22) but equatorial when R' is axial (23). This orientation would provide NMR chemical shifts δ_A and δ_E for appropriate nuclei in 22 and 23. When R' = H and equilibrium is thus rapid between 22 and 23, the system could then be probed by measuring the shifts of the same nuclei (δ) in the equilibrated system and using the equation we had established in 1959,[38] $\delta = n_A \delta_A + n_E \delta_E$, in which $n_A + n_E = 1$, and solving for n_E and n_A.

As with so many problems, success in this one depended very much on the ability of the co-worker. The first postdoctoral collaborator to whom I assigned the problem (immediately after my arrival in Chapel Hill) floundered; the synthesis of the required substituted *trans*-decahydroquinolines proved too difficult. Fortunately, Fritz Vierhapper from the University of Vienna joined my group in 1973. Just before leaving Notre Dame, I had had another very gifted postdoctoral collaborator from Vienna—Otmar Hofer, who had worked both on a conformational[84] problem and on a mechanistic[85] one with much success, and so I was pleased to have another person from the same university come to my laboratory.

Vierhapper observed that the hydrogenation of quinolines (and isoquinolines) in acid medium proceeded in two stages, the eventual result being the formation of *cis*-decahydroquinolines, as had been reported in the literature. Fritz, however, had the good sense to interrupt the hydrogenation after absorption of only two molecules of hydrogen and was rewarded by finding that the product was nearly all 5,6,7,8-tetrahydroquinoline; that is, the benzene ring was reduced first.[86] Under other conditions reported in the literature, almost invariably the pyridine ring is reduced first. Further reduction of the pyridine ring chemically (with sodium and ethanol) made it possible to produce *trans*-decahydroquinoline as the final product.[87] With this discovery, the synthesis of the substrates shown in Scheme 23 (22 and 23, R = CH_3) became straightforward, and ^{13}C NMR spectroscopy soon showed[80,88] that the $-\Delta G^\circ$ value for the N–CH_3 group was around 2 kcal/mol, which is at the upper end of the previously established range. (Later, the value for the same group in N-methylpiperidine was found[89] to be even larger, between 2.4 and 3.15 kcal/mol, depending on solvent and phase).

Vierhapper, largely on his own, carried out a number of additional studies with the decahydroquinoline system,[90,91] of which perhaps the most important one was the establishment of the N–H equilibrium[92] (Scheme 23; R = H). To accomplish this feat, he had to use the *tert*-butyl group at C-8 in *trans*-decahydroquinoline [Scheme 23; R' = C(CH$_3$)$_3$] as a biasing group to force the N–H into a predominantly axial position (conformation 22) and to effect the analysis by Bohlmann bands in the infrared spectrum rather than by NMR spectroscopy. We concluded that N–H also prefers the equatorial position (Scheme 23; 21, R = H) by about 0.5 kcal/mol in CCl_4. In effect, we came down on Katritzky's side (*see* page 75). We had to establish in this work that **23** [R' = C(CH$_3$)$_3$ and R = H] actually existed in a double-chair conformation with an axial *tert*-butyl group. This was confirmed by X-ray structure analysis[93] by Karl Hargrave, a very able crystallographer who had joined my group after receiving his Ph.D. degree with Andrew McPhail at Duke University.

^{13}C NMR Spectroscopy of Heterocyclic Compounds

During 1975–1982, we effected, by low-temperature ^{13}C NMR spectroscopy, the conformational analysis of a number of substituted saturated heterocyclic compounds: thianes,[94] piperidines,[95,96] oxanes,[97] thianium salts,[98] and piperidinium salts.[96,99] The methodology is quite simple in principle (Scheme 24, part a).[94] The ^{13}C NMR spectrum of the appropriate compound is recorded at −80 to −100 °C. Because this temperature is below the coalescence temperature, the signals of the two conformers (**24a** and **24b**) can be seen separately and integrated to give K = [24a]/[24b] and, hence, ΔG_a^o ($\Delta G_a^o = RT \ln K$). Thus for a methyl group at C-2 in thiane (Scheme 24, part a), $\Delta G^o = 1.42 \pm 0.07$ kcal/mol. However, the ratio [24b]/[24a] in this case is 41 ± 7, which is near the upper limit measurable by NMR techniques; $-\Delta G^o$ values greater than 1.4

24a   **24b**   (a)

$$-\Delta G_a^o \; = \; -\Delta G_{2\text{-Me}}^o = 1.42 \text{ kcal/mol}$$

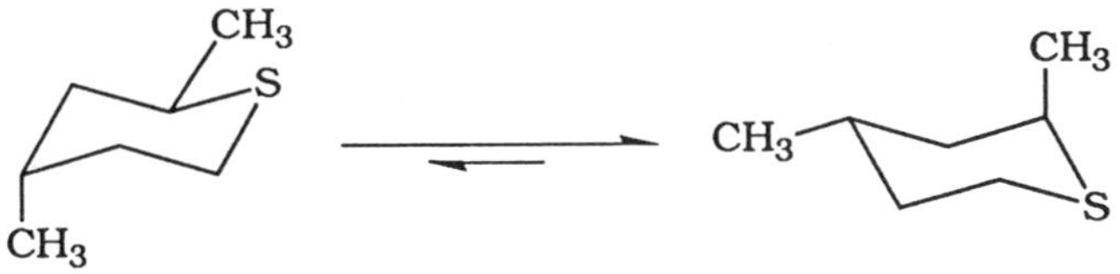

25a   **25b**   (b)

$$-\Delta G_b^o \; = \; -\Delta G_{4\text{-Me}}^o - (-\Delta G_{2\text{-Me}}^o) \; = \; 0.38 \text{ kcal/mol}$$

$$\text{Hence } -\Delta G_{4\text{-Me}}^o = -\Delta G_b^o - \Delta G_{2\text{-Me}}^o = 0.38 + 1.42 \text{ kcal/mol}$$

Scheme 24, *continued on next page*

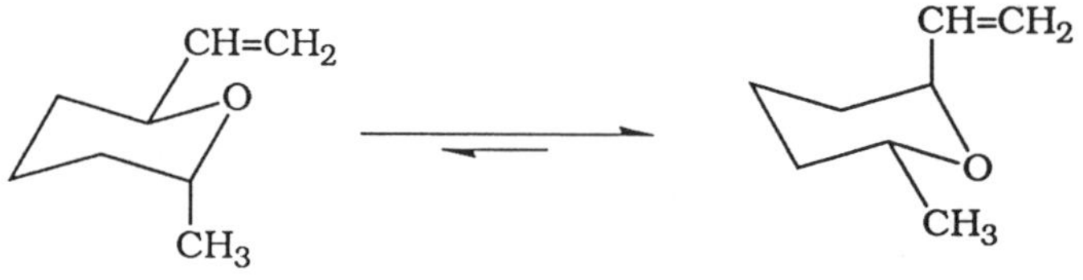

(c)

26a   **26b**

$$-\Delta G_c^o = -\Delta G_{2\text{-Me}}^o - (-\Delta G_{2\text{-vinyl}}^o) = 0.42 \text{ kcal/mol}$$

(d)

27a   **27b**

$$-\Delta G_d^o = -\Delta G_{2\text{-vinyl}}^o - (-\Delta G_{3\text{-Me}}^o) = 0.84 \text{ kcal/mol}$$

Summing: $-\Delta G_c^o - \Delta G_d^o = -\Delta G_{2\text{-Me}}^o - (-\Delta G_{3\text{-Me}}^o)$

$$-\Delta G_{2\text{-Me}}^o = -\Delta G_c^o - \Delta G_d^o - \Delta G_{3\text{-Me}}^o$$

$$= 0.42 + 0.84 + 1.43 = 2.69 \text{ kcal/mol}$$

Scheme 24, *continued*

kcal/mol ($K > 50$ at -80 °C) cannot be measured directly. Therefore, larger $-\Delta G^o$ values were measured in disubstituted compounds by assuming additivity of ΔG^o for the two substituents. (In cases in which independent measurements of ΔG^o for the two individual substituents were possible, additivity was generally confirmed.)

An example, $-\Delta G^o$ for 4-methylthiane, is shown in Scheme 24, part b. The ratio [25b]/[25a] (2.92 ± 0.17) is readily measurable, because the free energy gain from shifting the methyl group at C-4 from axial to equatorial is partially compensated by the movement of the methyl substituent at C-2 from the equatorial to the axial position. Thus $-\Delta G_b^o$ is readily measured, and by additivity (*see* bottom of Scheme 24), $-\Delta G_{4\text{-Me}}^o$ is computed from $-\Delta G_b^o$ and $-\Delta G_{2\text{-Me}}^o$ as 1.80 ± 0.10 kcal/mol, a value that is too large to be measured directly.

We have called this technique the counterpoise method,[79a] because the methyl at C-2 is used as a counterpoise to the methyl at C-4. Occasionally a "relay" is needed. For example, for 2-methyloxane, $-\Delta G^\circ = 2.86$ kcal/mol. This value is so large that counterpoising with a methyl group at C-3 ($-\Delta G^\circ = 1.43$ kcal/mol by direct equilibration), as in *cis*-2,5-dimethyloxane, is inadequate. Here a vinyl group at C-2 was used as a relay (Scheme 24, parts c and d: 26a, 26b, 27a, and 27b).[97a]

A summary of $-\Delta G^\circ$ values for methyl groups at C-2, C-3, and C-4 in various saturated six-membered heterocyclic compounds determined in our laboratory is given in Table II. The values for the methyl

Table II. $-\Delta G^\circ$ Values for C-2, C-3, and C-4 Methyl Groups of Six-Membered Saturated Heterocyclic Compounds

Compound (heteroatom)	C-2	C-3	C-4
Cyclohexane[a]	1.74	1.74	1.74
Thiane (S)	1.42	1.40	1.80
Oxane (O)	2.86[b]	1.43	1.95
Piperidine (NH)	2.5	1.6	1.9
N-Methylpiperidine (NCH$_3$)	1.7	1.6	1.8
N-Methylpiperidinium[c] (NDCH$_3$[+])	1.4	2.2	1.6

NOTE: Values are expressed in kilocalories per mole.
[a] Cyclohexane is included for comparison. *See* ref. 159.
[b] For the reason for the difference between this value and that given in Scheme 24, *see* ref. 97a.
[c] This compound was used as the hydrochloride in D_2O.

group at C-4 (which is remote from the heteroatom) vary little; what variation there is is probably due to differences in puckering of the different rings. The values at C-3 are uniformly smaller for the heterocyclic compounds than for cyclohexane, a result supporting the earlier mentioned (*see* page 56) conclusion that a *syn*-axial lone pair provides less van der Waals repulsion than does a *syn*-axial hydrogen atom. The value at C-3 is smallest for oxane and largest for piperidine and N-methylpiperidine. The value at C-2 is largest in oxane, followed by piperidine, cyclohexane, and thiane. This order logically relates to the distance between the *syn*-axial C-2 methyl and C-6 hydrogen ligands; the shorter this distance, the larger is $-\Delta G^\circ$. In turn, this distance relates to the carbon–heteroatom (or carbon–carbon, in cyclohexane) bond length; the shorter this bond length (C–S, 1.81 Å; C–C, 1.53 Å; C–N, 1.47 Å; C–O, 1.44 Å), the shorter is the CH$_3$(2a)–H(6a) distance, and the larger is $-\Delta G^\circ$.

There are two puzzles in this correlation, however: the relatively small value for $-\Delta G^{\circ}_{2-Me}$ in 2,N-dimethylpiperidine and the even smaller value for the corresponding salt. We believe that the anomalous value for N-methylpiperidine is due to a strong repulsive interaction of the N-methyl and C-2-methyl groups when both are equatorial (the origin of this effect has not been fathomed yet). The anomaly for the corresponding salt may be enhanced by a lengthening of the $C-N^{+}$ bond (to about 1.51 Å), which reduces the $CH_3(2a)-H(6a)$ *syn*-axial repulsion, as well as, possibly, by solvation. Solvation, occurring largely from the axial side of the nitrogen (the $N-CH_3$ being equatorial), would place the associated solvent *anti* to an axial C-2 methyl but *gauche* to an equatorial C-2 methyl.

The case of the piperidinium salts (Scheme 25) is particularly interesting for two reasons. First, at the appropriate pH, equilibrium can be established and studied by ^{13}C NMR spectroscopy at room temperature in D_2O solution. Although the salts do not equilibrate as such, deprotonation and reprotonation occur rapidly enough at pH 4–5 to allow equilibration via the (fast-interconverting) free amines but without converting an appreciable stoichiometric amount of the salt into free amine and without producing a coalescence of the N-methyl groups of the axial and equatorial species in the NMR spectrum. Second, and

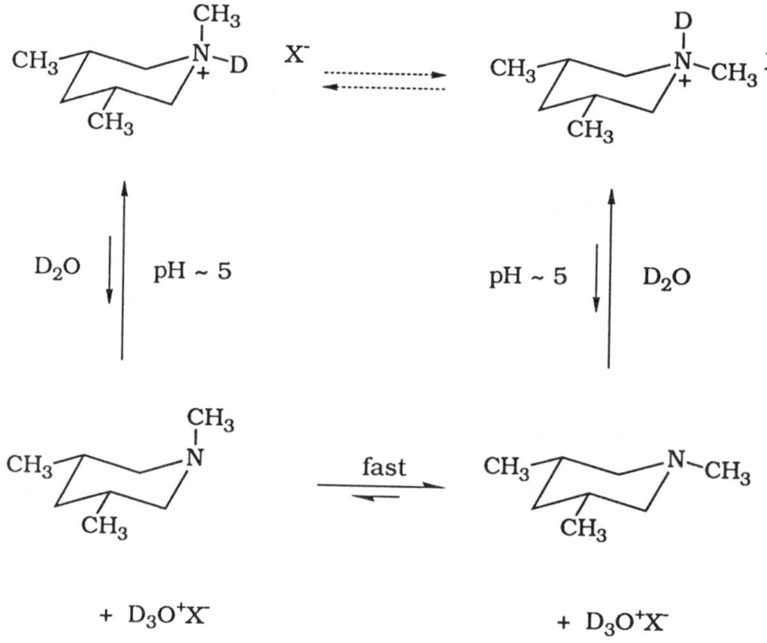

Scheme 25

this observation initially surprised us, the equilibrium of the salts in this case is also less on the side of equatorial N-methyl than the corresponding equilibrium of the free amines. Again, we ascribe this result to solvation: the axial N-methyl salt, with equatorial H or D, is somewhat better solvated from the equatorial side than is the equatorial N-methyl compound from the side of the axial H or D. This difference in conformation between free amines and their salts is often overlooked by pharmacologists.

The availability of the various heterocyclic compounds also allowed us to study and correlate their ^{13}C NMR spectra.[94,96,100] Later we carried out related studies with ^{17}O NMR spectroscopy.[101] The ^{13}C NMR spectroscopic work has been reviewed[102] in collaboration with K. Michal ("Mike") Pietrusiewicz, an able postdoctoral collaborator from the Centre for Molecular and Macromolecular Chemistry in Lodz, Poland, who also contributed actively to our experimental studies with ^{13}C NMR spectroscopy.

1974—The Centennial of Stereochemistry

Before describing our work during the last 12 years (1977–1989), I should like to go back to the year 1974—an important year for stereochemistry because it marks the centennial of the discovery of the concept of the "asymmetric carbon atom" by Le Bel and van't Hoff. A number of international meetings (chronicled in more detail elsewhere[103]) took place that year, in several of which I participated and one of which I helped organize. The first of these meetings was the van't Hoff Centenary held in Leiden, the Netherlands, in early May of 1974 and organized by my friend and colleague Egbert Havinga, then a professor of organic chemistry at Leiden and himself a well-known stereochemist. I have three vivid memories of this meeting: we were given replicas of van't Hoff's original models, as well as of his original 1874 publication (published as a pamphlet in the Dutch language), we were treated to a lecture by Bill Lipscomb (of Harvard University) on carboxypeptidase, which was illustrated with impressive three-dimensional projections (we all wore polarized eyeglasses), and, sadly, I listened to Oosterhoff's last lecture (he died soon thereafter and was already quite ill at the Congress), which included a vivid description of the "Tanaka affair".[103] In 1972, J. Tanaka had claimed that Bijvoet's assignment of the absolute configuration of tartaric acid needed to be reversed. The claim had been invalidated subsequently by an ingenious scattering technique by H. H. Brongersma (who is Oosterhoff's son-in-law). The technique showed that the Bijvoet experiment did, in fact, give the correct result.

In Leiden, Eva and I stayed at a very charming family hotel, the Witte Singel, where Havinga had reserved a room for us. Also at the same hotel was the Swedish stereochemist Arne Fredga, who is well known for his work on the determination of configuration by using quasi-racemates. Fredga was known to be a member of the Nobel Committee for the prize in chemistry. Because the new chemistry building in Leiden was quite a distance from the downtown hotel, he and I shared a taxi one morning, and I took the opportunity to invite him to the van't Hoff—Le Bel Centennial Symposium in Atlantic City, NJ, to take place in September of 1974. Fredga responded that he could come if the chemistry Nobel laureate for 1974 had been decided by the time of the Atlantic City meeting; otherwise he would have to stay in Stockholm. In my usual brashness, I allowed that the decision for 1974 should be easy; the obvious champions were Woodward and Hoffmann. Whereupon Fredga reminded me that Woodward already had a Nobel prize in chemistry. My response was that, after all, Bardeen had already had a Nobel prize in physics when he got a second one. Fredga's reply was that Bardeen had received one-third of the prize originally and later on received one-half, so the total was only five-sixths. But Woodward had received the prize all for himself, and one whole prize in one

At the 1974 Bürgenstock Conference with Harry Mosher (right). Note the hungry look!

Ernest L. Eliel (far right) at the Taverna in Bürgenstock in 1974. Left side, left to right: Nathan and Rachel Kornblum, the late Franz Sondheimer, a guest, and Phil Eaton.

subject, according to Fredga, was all one could get! (I am not sure whether this statement represents the policy of the Nobel Committee or just Fredga's personal opinion; the fact is that Roald Hoffmann received the prize, jointly with Fukui, only after Woodward's death.)

From Leiden we proceeded to the 9th Bürgenstock conference, and from there we went on to Spain, where I gave a short course on conformational analysis at the Universidad Complutense in Madrid, in Spanish. After that, I lectured in the south of Spain, and Eva, my younger daughter Carol, who had joined us, and I did some sightseeing in Seville, Cordoba, and Granada. The total trip to Europe had lasted nearly 6 weeks (even before the Leiden meeting I had attended my nephew's wedding in London and then a physical–organic chemistry symposium in Noordwijkerhout in the Netherlands). Immediately after our return, we attended my older daughter Ruth's graduation at Smith College. When we returned to Chapel Hill, I had a very high fever and was quickly diagnosed as having hepatitis. Fortunately, it was not of the usual viral origin. A very thorough resident at our hospital proved unequivocally that the liver infection was due to mononucleosis, a very unusual disease for a 53-year-old person!

I spent the next 5 weeks at home in bed; I still remember Fritz Vierhapper coming to our house and explaining his latest results by drawing on an artist's pad (I had no blackboard at home). It was, touch wood, one of the rare occasions in my professional life when I was seriously ill, but it came at a bad time, because I still had two important responsibilities to fulfill that summer: first, to run the van't Hoff–Le Bel Centennial Symposium at the fall meeting of the American Chemical Society in early September and, second, to attend the van't Hoff–Le Bel Centennial and give a plenary lecture (in French) in Paris right after the Atlantic City meeting.

The Atlantic City meeting involved not only technical presentations but also historical lectures,[104] which were organized by Bertrand Ramsay of Eastern Michigan University. The technical lectures were given by Arigoni, Havinga, Hoffmann, Lehn, Lipscomb, and Mislow; the historical lectures included one by Prelog. The Russian historian of stereochemistry, G. V. Bykov, was also on the program but, as then so often happened with Soviet scientists, was unable to attend; his lecture had to be read. Perhaps the most unusual part of the symposium was a dinner followed by a play. To the dinner I had invited the Dutch chemist P. H. Hermans, who, unbeknownst to either Winstein or myself, had virtually anticipated[105] our equation[29–31] on the kinetic behavior of conformationally mobile systems in 1924! Unfortunately, Hermans' work had been widely overlooked, and I myself had come upon it only shortly before, in the course of preparing a lecture to be presented at an ACS meeting before the Division of Chemical Education.

The circumstances are of some interest: Hermans was a student of Böeseken in Delft; Böeseken's pioneering stereochemical work is well known and widely cited. Apparently, Hermans, a thorough and quantitative thinker, did not get along with Böeseken, and as a result, he published his thesis work by himself in a physical chemistry journal in German[105] and he failed to obtain an academic post and spent the rest of his professional life in industry. Thus, Hermans was unable to pursue his pioneering work on the conformational analysis of acyclic systems.

From earlier correspondence, it was clear to me that Hermans was understandably unhappy about never having received credit for this work. Therefore, it was a great pleasure to both of us that, with the help of my local colleague J. J. Hermans (no relative) and the Dutch chemical industry, he was able to come to the United States for the 1974 meeting. At the dinner, I said a few words about his early contributions. When he rose to respond, his voice was almost choked. Others later told me how much this recognition meant to him. It is a continuing source of satisfaction to me that ultimately—and while he was still alive—P. H. Hermans received the recognition due to him. Of course, no one shall ever know what additional contributions to conformational analysis he might have made if he had been given the opportunity. (The same is true for another pioneer of conformational analysis, Arnold

Weissberger, who, at a critical stage in his career, was forced out of his academic position in Germany by the Nazis).

After the dinner, George Kauffman (of California State University), a participant in the historical part of the symposium, staged a performance of "Rotating and Resolving" (*Drehen und Spalten*), a play purportedly written by the famous inorganic stereochemist and Nobel laureate Alfred Werner (of the University of Zurich) during the early years of the 20th century. Kauffman had arranged for a noteworthy cast—from Vlado Prelog to the then editor of *Chemical and Engineering News*, Albert Plant. The play was very well received by the audience. From Atlantic City, I proceeded directly to Paris to give the Le Bel Centennial lecture in French; but my French on this occasion was not as fluent as it had been in 1968.

A New Venture: Enantioselective Synthesis

I mentioned before that the very high stereoselectivity of the 1,3-dithiane reaction Hartmann had discovered (Scheme 21) suggested to us in the early 1970s that it might be developed into an asymmetric (or, more properly, enantioselective) synthesis. In a contribution[74] to the 1974 van't Hoff–Le Bel memorial issue of *Tetrahedron*, I mentioned this possibility and laid down the rules for a viable enantioselective synthesis, but it took us 3 more years and another chance discovery to transform the idea into practice.[106] This delay was unfortunate, because

A break from the 1980 Gordon Conference on Stereochemistry with former students and postdoctoral fellows. Left to right: F. Vierhapper, K. Soai, William Bailey, Ernest L. Eliel, and E. Juaristi.

whereas in the early 1970s enantioselective syntheses of high selectivity were virtually unknown (H. C. Brown's asymmetric hydroboration[107] being a notable exception), by the mid-1970s, under the impetus of discoveries of A. I. Meyers,[108a] W. Knowles,[108b] and others, these syntheses became more common. By the early 1980s, when we began to exploit fully our own enantioselective synthesis, they had become commonplace.[109]

By way of an amusing sidelight, the definitive book on asymmetric synthesis in the early 1970s was written by Morrison and Mosher,[110] in part while Harry Mosher was in Zurich in 1967–1968, when we were both NSF senior postdoctoral fellows and lived in the same apartment building. However, although the book was thorough and extensive, it contained hardly any enantioselective processes of practical importance. Indeed, the initial interest in the book was sufficiently slack for the publisher to let it go out of print after only 3 years. At that time I was active in the ACS Committee on Publications, and it was largely thanks to my efforts—and over the objections of the then executive director of the ACS—that the ACS reprinted the book in 1976. By that time—and perhaps partly stimulated by the appearance of the book—there was great interest in the topic of asymmetric (in the sense of enantioselective) synthesis. I am told that many more copies of the reprint of the book have been sold than of the original edition!

Let me now go into our enantioselective synthesis in some detail. The 1,3-dithiane shown in Scheme 21 is not chiral, but even if it were (cf. Scheme 26; **28**, X = S), the transfer of chirality from C-6 to C-2 would not be fruitful, because one of the requisites of an enantioselective synthesis[74] is that the newly created chiral center (in this case, C-2) can be cleanly (and without loss of stereochemical integrity) separated from the original (auxiliary) one (in this case, C-6). Clearly, this requirement cannot be satisfied by a 1,3-dithiane (**29**, X = S). However, it occurred to us that a 1,3-oxathiane (Scheme 26; **28** and **29**, X = O) might satisfy this requirement. First, however, we had to demonstrate that 1,3-oxathianes, like 1,3-dithianes, can be alkylated by the Corey–Seebach procedure and that the alkylation (like that shown in Scheme 21) leads entirely to equatorially substituted products.

Breakthrough in Oxathiane Synthesis

This work was done[111] (Scheme 27) by Jorma Koskimies, a Finnish graduate student who had moved with me, reluctantly, from Notre Dame to Chapel Hill (he claimed that Chapel Hill was too hot for someone brought up in the climate of Finland!). We found that oxathianes substituted in positions 4, 6, or both (although, surprisingly, not the parent

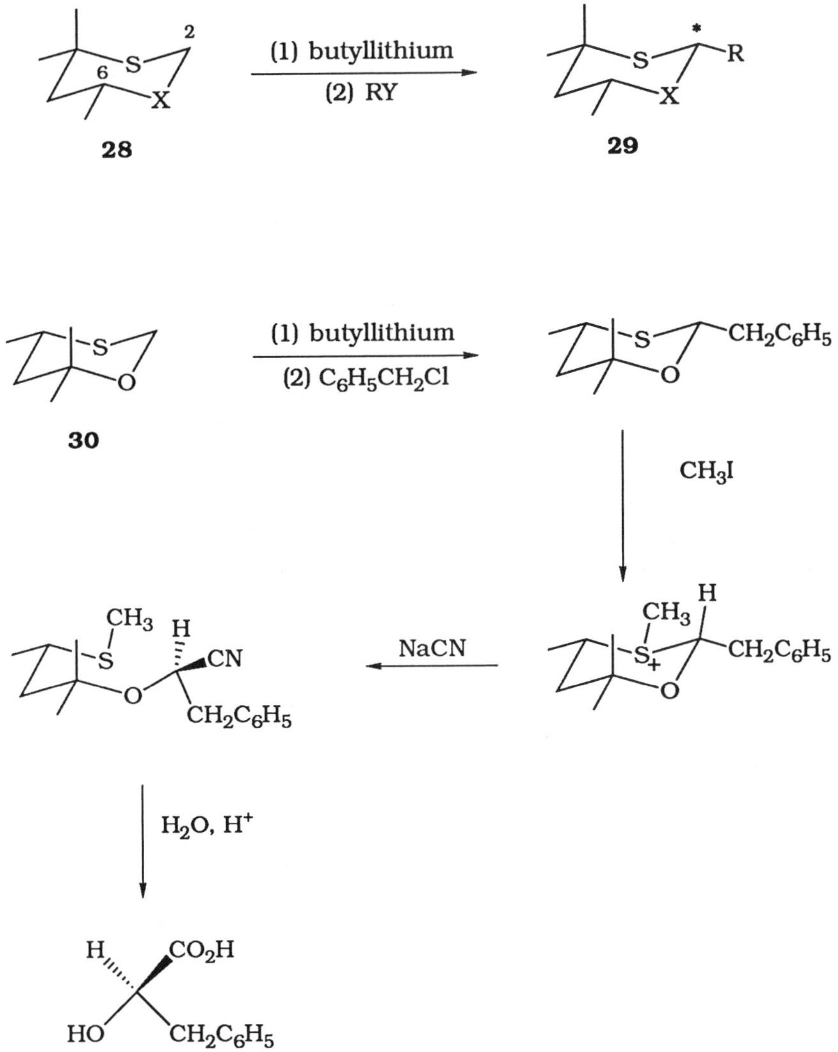

Scheme 26

compound) form lithium derivatives quite readily with butyllithium and that these lithium derivatives yield virtually exclusively equatorial products in electrophilic substitution. (It is best to manipulate the lithium derivatives at −25 °C because they are somewhat less stable than the 2-lithio-1,3-dithianes shown in Scheme 26, **28** and **29**, X = S.)

However, when we resolved substrate **30** (similar to **28**, X = O, in Scheme 26) and converted it to (S)-α-phenyllactic acid by the route shown in Scheme 26 (bottom), the product was substantially racemized.

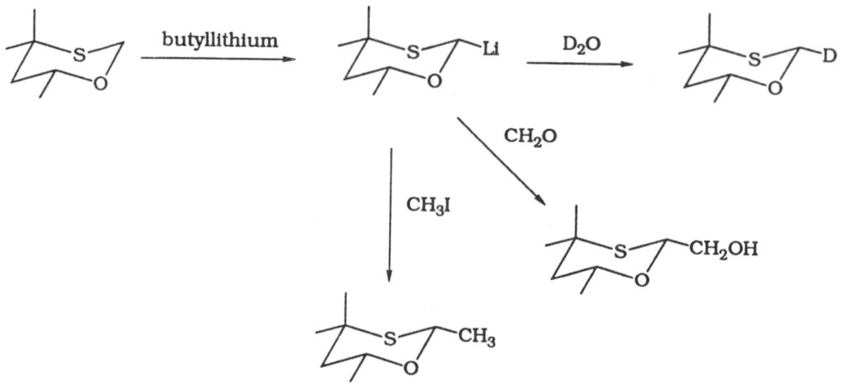

Scheme 27

Fortunately, just at this moment, a better approach was discovered by accident. To check further the generality of equatorial substitution in oxathianes (Scheme 27), Koskimies had studied the reaction with both acetophenone and ethyl benzoate (Scheme 28); in both cases, NMR spectroscopy showed that the products were entirely equatorial, but in the case of acetophenone, two epimers (at the exocyclic chiral center) were obtained as expected.

We could have rested our case here, but I suggested to Koskimies that, as a matter of good form, he should add methylmagnesium iodide to the ketone resulting from the ethyl benzoate reaction and make sure that he got the same carbinols as those from the acetophenone addition. Koskimies found that it was indeed so, but the product was almost exclusively (>99%) one of the diastereomers at the exocyclic chiral center (Scheme 28). We quickly realized that we were seeing an extreme example of the operation of Cram's chelate rule[112] and that by hydrolyzing the product we should be able to isolate a virtually enantiomerically pure α-hydroxyaldehyde RR'C(OH)CHO.

The realization of this scheme was not trivial and required the help of an excellent Swiss postdoctoral collaborator, Bruno Lohri, who came in 1975 from Hardegger's group at the ETH in Zurich. It was, of course, necessary to resolve the precursor 4,4,6-trimethyl-1,3-oxathiane (Schemes 27 and 28) or the equivalent 4,6,6-trimethyl-1,3-oxathiane (Scheme 26, 30). Optically active 4,4,6-trimethyl-1,3-oxathiane was obtained by Fritz Vierhapper through the asymmetric reduction of $C_6H_5CH_2SC(CH_3)_2CH_2COCH_3$ [the addition product of $C_6H_5CH_2SH$ and $(CH_3)_2C=CHCOCH_3$] followed by debenzylation (with Na and liquid NH_3) and treatment with paraformaldehyde and acid (by using the conditions worked out by Koskimies for the racemic material). The synthesis of 30 was achieved by Lohri through the resolution of

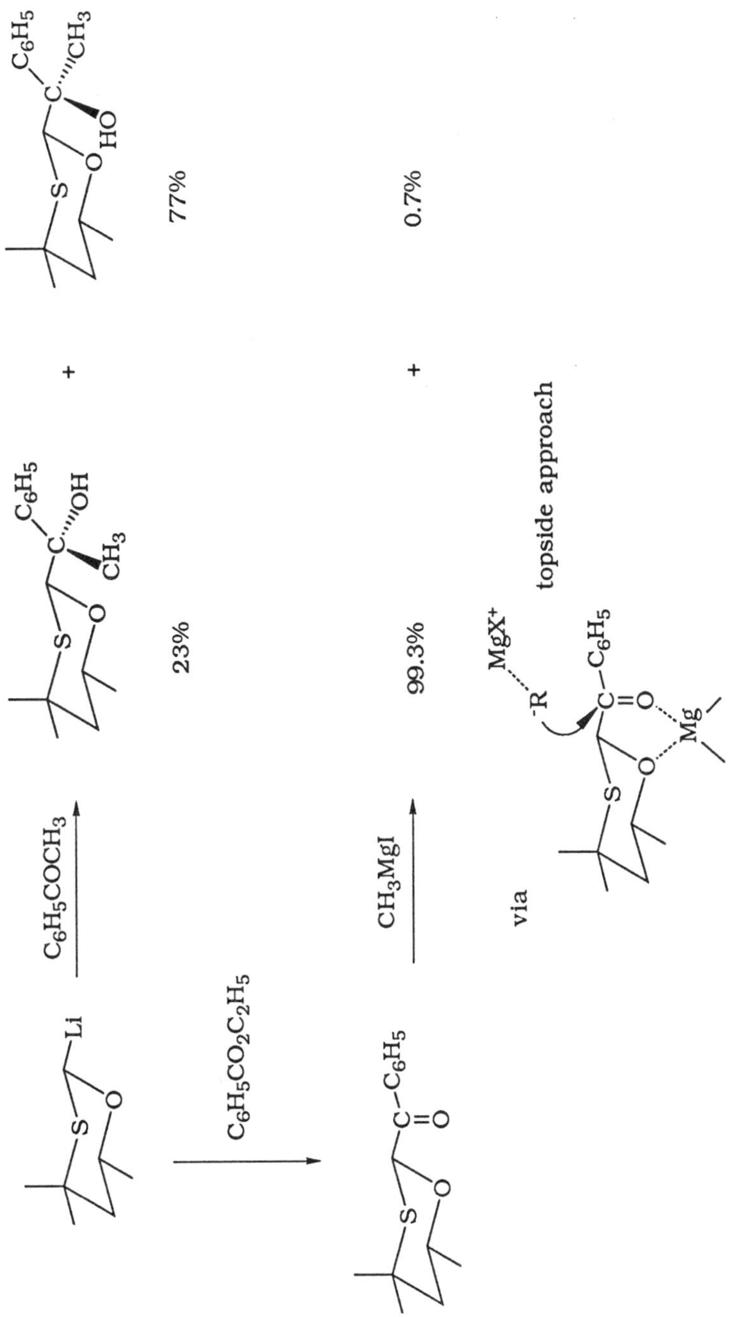

Scheme 28

$C_6H_5CH_2SCH(CH_3)CH_2CO_2H$ followed by esterification, treatment with CH_3MgI, debenzylation and treatment with dimethylformal (2,2-dimethoxymethane) and acid. Unfortunately, resolution of the acid by the method described in the literature (by crystallization of the cinchonidine salt) is quite tedious in requiring nine crystallizations. We, therefore, contented ourselves in working with acid of 44% enantiomeric excess; this decision had some unfortunate consequences, soon to be detailed.

Lohri encountered (and overcame) several additional serious problems. One difficulty was that the condensation with ethyl benzoate (Scheme 28, left) proceeded with unacceptably poor yield. Lohri therefore chose, instead, reaction with benzaldehyde to give the corresponding carbinol (in good yield) followed by oxidation. This was not as easy as we had expected, because most oxidations not only converted the alcohol to ketone but also the sulfide moiety to sulfoxide or sulfone. Lohri finally found that the Swern method[113] [oxidation with DMSO, $(CF_3CO)_2O$ and $(C_2H_5)_3N$] worked selectively. The oxidation reaction was followed by a mild cleavage with $CH_3I-CaCO_3-CH_3CN$ to give atrolactaldehyde methyl ether, which was further oxidized to atrolactic acid methyl ether (Scheme 29) with a 44% enantiomeric excess,

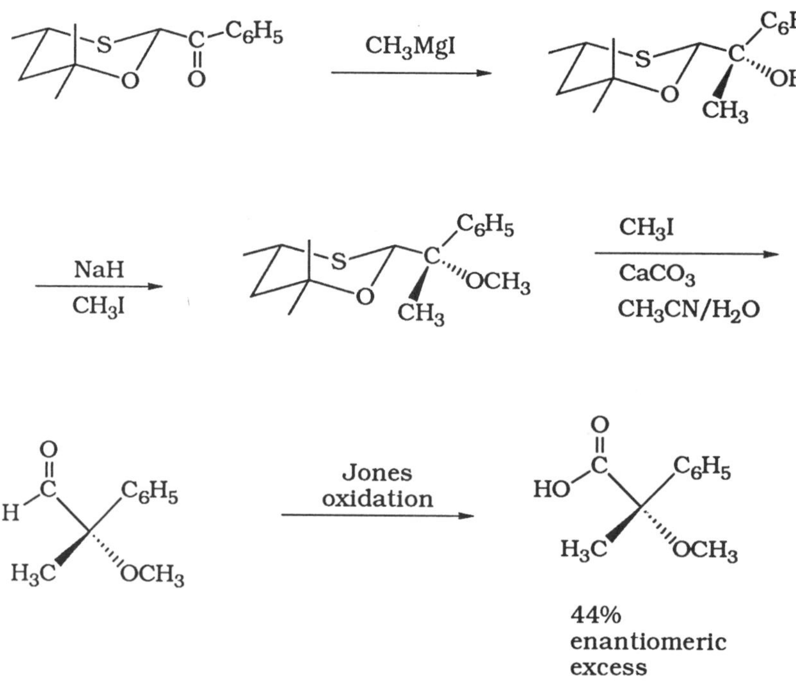

Scheme 29

which was, within the limits of experimental error, the same as that of the starting oxathiane. This result means that the optical yield in the synthesis was near 100%.

Incidentally, this asymmetric synthesis achieved by Lohri and Koskimies[106] illustrates some of the advantages afforded by our proximity to and good relations with the research laboratories of the Research Triangle Park, about 12 miles from the University of North Carolina. In September 1977, there was an IUPAC meeting in Tokyo that I was anxious to attend (I had never been in Japan before). I had been invited to give a lecture at this meeting, and I had taken the unusual step (for me) of proposing as the topic of the lecture research results (in this case, results from our asymmetric synthesis) that, at the time of submission of the abstract, were not yet complete.

Unfortunately, and to Lohri's despair, our polarimeter broke down a few weeks before the synthesis was finished. Before Lohri could proceed to the next step in his synthetic scheme, he had to measure the rotation of the intermediate he had brought up. (Although we never use optical rotation as an index of enantiomeric excess [ee] any more, it does serve as an easy-to-use analytical control in the course of an asymmetric synthesis.) Fortunately our friends at the Burroughs Wellcome Research Laboratories (for whom I am also a consultant) were willing to arrange for Lohri to use their polarimeter every afternoon, so

"100% enantiomeric excess!" Ernest L. Eliel at his 1979 birthday party.

he could check his products before proceeding to the next step of the synthesis.

The lecture in Tokyo was an unqualified success. It was attended by several hundred persons. The next year (May–June 1978), I returned to Japan under the sponsorship of the Japan Society for the Promotion of Science and lectured from Kyushu to Hokkaido about the new synthesis. The visit was wonderful from the personal, as well as

Masaru Fukuyama, Genichi Tsuchihashi, Ernest L. Eliel, Noboru Kito, and Iwao Ojima in Sagami Labs in Japan in 1978.

scientific point of view, thanks in no small part to the hospitality and generosity of my Japanese hosts. Chief among them was Osamu Simamura, the sponsor of my visit, who had been a postdoctoral researcher in my group in 1960 and had become director of research of the Sagami Laboratory after retiring from his position as a professor of organic chemistry at the University of Tokyo.

The written account of the synthesis is another story. The report was submitted as a communication to the editor of the *Journal of the American Chemical Society* early in 1978.[106] One referee (Harry Mosher, who signed his report) recommended acceptance with minor modifications. The other referee was surprisingly balky. He objected to the product being only 44% enantiomerically pure and did not want to accept our argument that because $C_6H_5CH_2SCH(CH_3)CH_2CO_2H$ (*vide supra*) had been resolved completely, it followed that a synthesis of

At a tutorial at Tohoku University with Y. Senda (seated, left) and students in 1978.

product in 100% ee was, in principle, possible. He objected to our not having recovered the chiral auxiliary substance, and surprisingly, he complained that we had studied additions to phenyl ketones (cf. Scheme 29) only. He said that such additions were normally more stereoselective than the corresponding additions to alkyl ketones. After some difficulty, we persuaded the cognizant associate editor (Barry Trost) to accept the communication. We had similar difficulties in getting the project funded by the NSF, but again, we succeeded in the end.

We needed, as rapidly as possible, to generalize the synthesis, to find a way to recover the chiral adjuvant, and to synthesize a chiral adjuvant that was 100% enantiomerically pure, preferably one derived from a readily available natural product. The first problem was tackled by Susan Morris-Natschke, who applied the stereoselective synthesis shown in Scheme 28 (bottom) to a wide variety of ketones and Grignard reagents in different solvents, such as ether and tetrahydrofuran (THF), under a variety of conditions.[114] She found that the synthesis was generally highly stereoselective, although in the case of alkyl ketones, she had to operate at −70 °C. As the churlish referee of our communication had anticipated, alkyl ketones, indeed, tend to react with nucleophiles less stereoselectively than do aryl ketones!

The cleavage reaction was ably perfected[115] by Joe Lynch, another graduate student. Lynch used *N*-chlorosuccinimide–silver

Ernest L. Eliel with Osamu Simamura at the Waterwheel Restaurant in Sagami, Japan, in 1983.

nitrate,[116] which not only cleanly produced the product hydroxyaldehyde RR'C(OH)CHO but also led to the recovery of the chiral auxiliary in the form of a sultine (Scheme 30). The chiral auxiliary was readily reconverted to the starting oxathiane by LiAlH$_4$ reduction followed by treatment with paraformaldehyde and acid. Thus one of the conditions laid down in 1974,[74] namely that the chiral auxiliary reagent must be recoverable in good yield and undiminished enantiomeric purity, was finally fulfilled.

Lynch also found that the free hydroxyaldehydes could be oxidized to the corresponding acid (or methyl ester or both) by iodine and KOH in methanol.[117a] Later we found sodium chlorite[117b] to be an even better selective oxidizing agent, because it leads exclusively to the free acid.

After several trials, the problem of synthesizing an optically active oxathiane from an available natural product had also been solved. Lynch's first attempt to start from a sugar was not successful, and a later synthesis of an optically active oxathiane from the two enantiomers of camphor-10-sulfonic acid[118] was not satisfactory in its applications. The

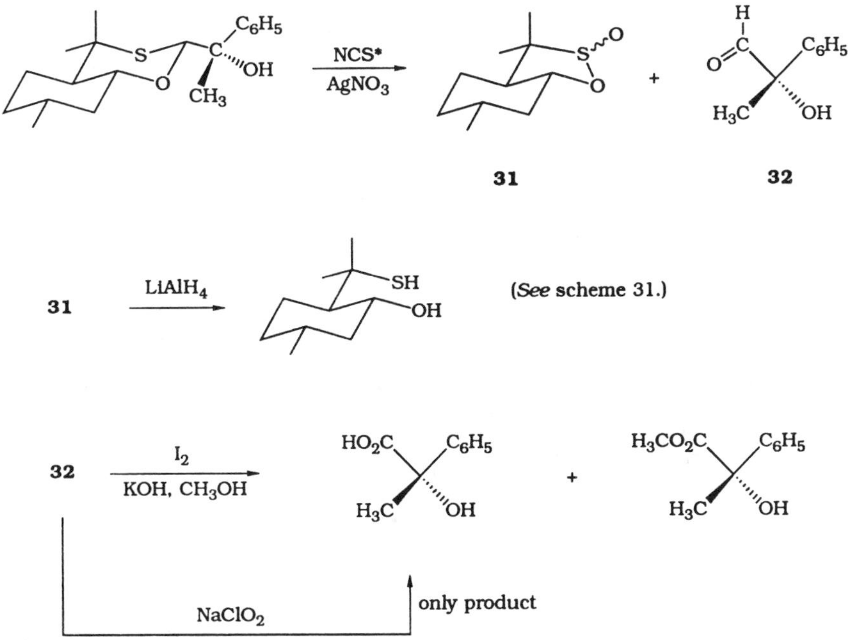

31   **32**

(*See* scheme 31.)

only product

* NCS is N-chlorosuccinimide.

Scheme 30

successful chiral auxiliary was eventually synthesized from pulegone (Scheme 31).[115,119]

The final product is crystalline and is thus readily freed of its diastereomers. Using this chiral template, Lynch synthesized various chiral tertiary α-hydroxyaldehydes [RR'C(OH)CHO] and, from these, glycols [RR'C(OH)CH$_2$OH], tertiary methylcarbinols [RR'C(OH)CH$_3$], and α-hydroxyacids [RR'C(OH)CO$_2$H]. Later, K.-Y. Ko, a Korean doctoral student, showed that ketones of the type used by Lynch can also be stereoselectively reduced to give rise to secondary α-hydroxyaldehydes (usually protected as O-benzyl derivatives) and their congeners (Scheme 32).[120]

Interestingly, Ko also found that DIBAL (diisobutylaluminum hydride), [(CH$_3$)$_2$CHCH$_2$]$_2$AlH, which is apparently incapable of chelating and therefore does not follow Cram's rule, gives the opposite configuration at the carbinol center from L-selectride. Several preliminary accounts of this work[121–123] were quickly published.

The configurations of several of the hydroxyacids and carbinols synthesized by Lynch and Ko (Schemes 30–32) were known in the literature, and so we were able to convince ourselves that Cram's

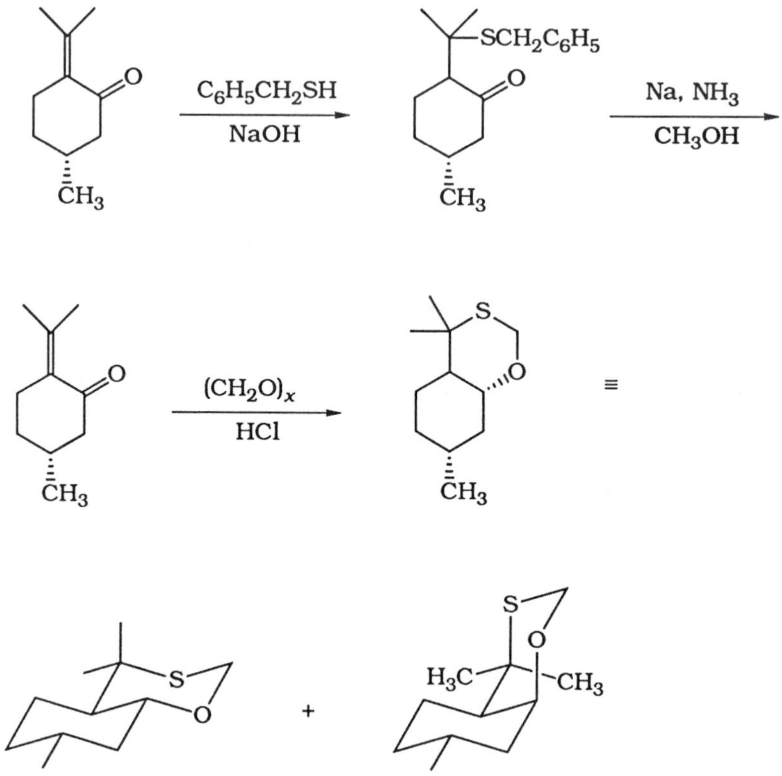

main product (crystalline)

isolable byproduct

Scheme 31

chelate rule was followed generally and that the synthesis could be used to assign configuration to the products in cases when it was not known.

Confluence of Empirical Rules

However, one apparent exception was (*S*)-(−)-α-*tert*-butyllactic acid (Scheme 33; **33**). On the basis of earlier work involving Prelog's rule,[124] the *S* configuration had been assigned to the dextrorotary acid. This acid is of some considerable importance, because it had been obtained by oxidation from and thus correlated with (−)-*tert*-butylethynylmethylcarbinol, which, in turn, had been related configurationally to (−)-1-chloro-3-*tert*-butyl-3-methylallene, $CHCl{=}C{=}C(CH_3)C(CH_3)_3$, one

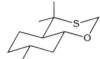
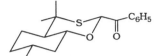

(1) butyllithium
(2) C$_6$H$_5$CHO
(3) Swern oxidation

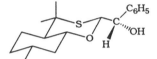

LiBH(CHCH$_2$CH$_3$)$_3$
|
CH$_3$

(L-Selectride)

HO$_2$C, C$_6$H$_5$

H   'OCH$_2$C$_6$H$_5$

(1) NaH, C$_6$H$_5$CH$_2$Cl
(2) NCS, Ag$^+$
(3) NaClO$_2$

Scheme 32

Ernest L. Eliel in July 1981 at the United States–Japan Conference on Asymmetric Synthesis at Stanford with co-organizer Sei Otsuka and wife Eva Eliel.

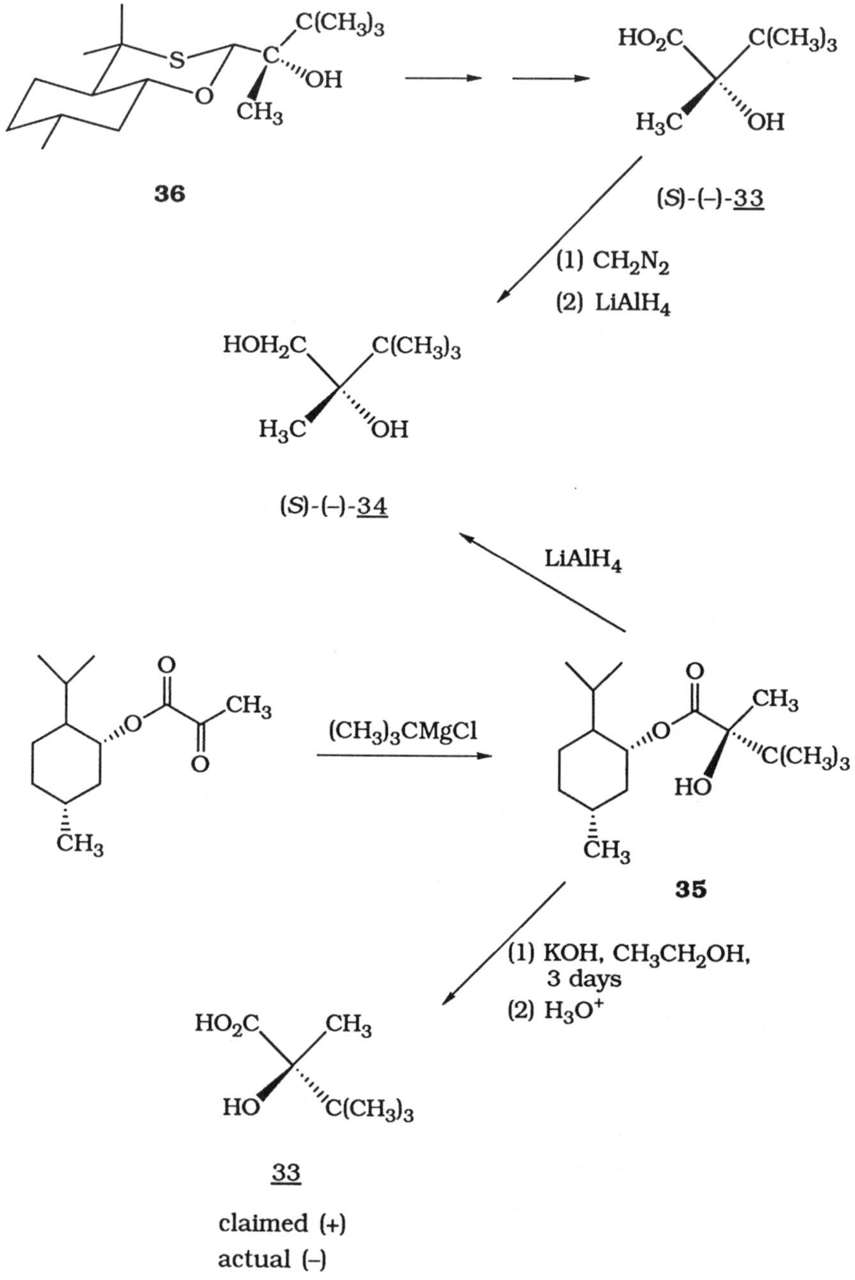

Scheme 33

of the first allenes whose absolute configuration had thus been established supposedly as S.[124]

We felt that although it was possible that Cram's chelate rule would break down in an addition to a *tert*-butyl ketone, an alternative possibility was that the earlier assignment[124] of the S configuration to $(+)$-α-*tert*-butyllactic acid was in error. In particular, we noted that the rotation of the acid was quite low, that the acid had been hard to purify, and that the saponification of the menthyl ester **35** was far from quantitative. Under these circumstances, as had already been pointed out by Prelog in the original publication[125] of "Prelog's rule", kinetic resolution may occur, that is, the minor product of the Grignard addition may be saponified faster and the acid thus obtained in major amount would correspond to the less abundantly formed ester (Scheme 33). This argument would clearly invalidate the original conclusion.

To convince ourselves that this hypothesis was to the point, Lynch quantitatively reduced the menthyl ester **35** (Scheme 33) with $LiAlH_4$ to the glycol **34**; this glycol, which has the S configuration according to Prelog's rule, indeed has the same (negative) rotation as the glycol obtained by the reduction of our acid **33** (Scheme 33). Moreover, when we saponified the menthyl ester **35** over a period of 3 days (to ensure nearly complete reaction and thus obviate any kinetic resolution) and compared the signs of rotation of both the acid and the corresponding methyl ester (the methyl ester has a much higher, but positive, specific rotation) with corresponding products of our oxathiane synthesis, it again became clear that both the Prelog and the Cram routes give the (S)-$(-)$-acid **33** and its (S)-$(+)$-ester; that is, that the two rules agree.

Ernest L. Eliel on the day he received a D.Sc. (H.C.) at Duke University, May 8, 1983.

At the same time that we were working on this problem, Stephen Baldwin, our neighbor at Duke University, had synthesized the same glycol (Scheme 33; 34) by Sharpless[126] oxidation of the allyl alcohol $CH_2=C[C(CH_3)_3]CH_2OH$ to the corresponding epoxide followed by hydride reduction. Again, Baldwin had convinced himself (on the basis of Sharpless's generalization of the steric course of epoxidation) that the (−)-carbinol had the S configuration. Therefore Prelog's rule, Cram's chelate rule, and Sharpless's rule agreed in leading to the same configurational assignment. Either all three rules give the correct assignment or all three rules must be wrong, at least in this case.

The fate of the resulting communication to the editor was extraordinary. When we submitted it in 1982, it was rejected on the basis of the report of a single referee, who felt that our assignment, resting as it did on the combined weight of Prelog's, Cram's, and Sharpless's rules, was inconclusive, because these rules are strictly empirical. Our argument that, whereas one rule might predict the wrong result, the chances of all three doing so were minuscule was not accepted.

We withdrew the communication in the hope that we could prove the configuration of the acid (Scheme 33; 33) by a crystallographic study of its precursor 36, which would correlate the configuration of the exocyclic (tertiary alcohol) chiral center with that of the oxathiane ring. The absolute configuration of the oxathiane is known by its provenance from (R)-(+)-pulegone (Scheme 31), whose absolute configuration is known. Unfortunately this project has not succeeded so far because of our inability to obtain suitable crystals either of the carbinol 36 (Scheme 33) or of a derivative.

As a result of our inability to publish this work, a fair amount of confusion, which was documented in a review[127] and which was also evident in original work,[128] ensued when others realized that the S configuration assigned to (−)-1-chloro-3-tert-butyl-3-methylallene (*vide supra*) was inconsistent with other data. Two papers finally appeared,[129] in which the configuration of the chloroallene was corrected to R on the basis of mechanistic studies. As a result, we resubmitted our communication with essentially the original arguments, and it was finally published in 1987![130]

Applications of Enantioselective Synthesis

During the last 5 years, we have published a number of applications of the oxathiane synthesis[131–138] (Scheme 34). Much of this work has been reviewed[139] and it will not be repeated in detail in this account. However, an interesting point embodied in the malyngolide synthesis[136] is

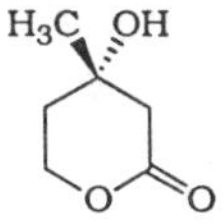

(S)-$(+)$- [and (R)-$(-)$-] mevalolactone

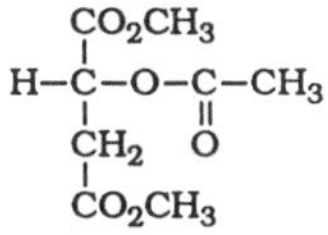

(R)-$(+)$- [and (S)-$(-)$-] dimethyl acetylcitramalate

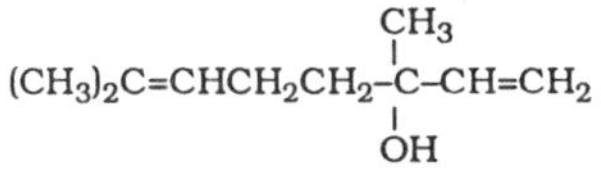

$(CH_3)_2C{=}CHCH_2CH_2{-}\overset{\overset{\textstyle CH_3}{|}}{\underset{\underset{\textstyle OH}{|}}{C}}{-}CH{=}CH_2$

(S)-$(+)$-linalool

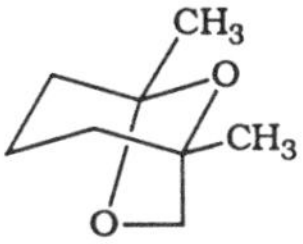

(S)-$(+)$- [and (R)-$(-)$-] frontalin

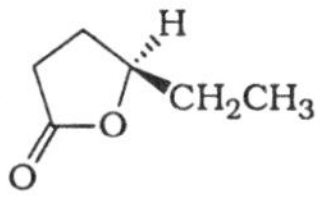

(R)-$(+)$- [and (S)-$(-)$-] γ-caprolactone

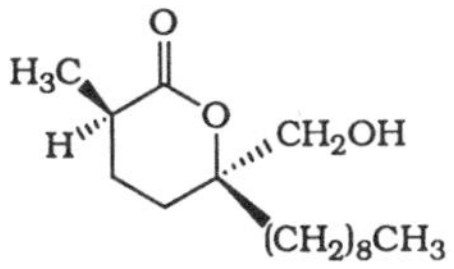

$(2R, 5S)$-$(-)$- [and $(2S, 5R)$-] $(+)$-malyngolide (and both diasteromers: $2R, 5R$ and $2S, 5S)$

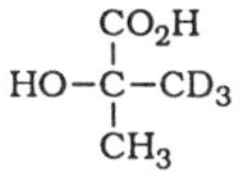

CO_2H
$HO{-}\overset{\overset{\textstyle }{|}}{\underset{\underset{\textstyle CH_3}{|}}{C}}{-}CD_3$

(R)-$(-)$- [and (S)-$(+)$-] β,β,β-trideuterio-α-hydroxyisobutyric acid

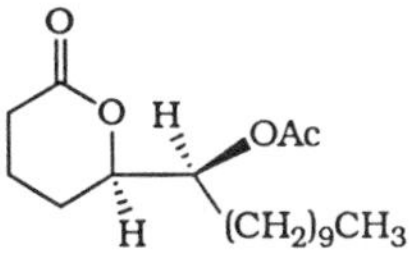

$(5R, 6S)$-$(-)$-6-acetoxy-5-hexadecanolide and its two enantiomeric diasteromers

(Both enantiomers were synthesized where shown, but only the first is depicted.)

Scheme 34

worth mentioning. As seen in Scheme 34, malyngolide has two chiral centers and therefore exists as four stereoisomers. Thanks to the experimental skill of Tetsuo Kogure, a Japanese postdoctoral researcher, all four stereoisomers were obtained in high enantiomeric purity by a convergent synthesis (Scheme 35), in which the two chiral centers (the C-2 methyl and the C-5 carbinol center) were independently and enantioselectively synthesized (in both *R* and *S* configurations) and put together to give the *RR, RS, SR,* and *SS* isomers. The percentage of the major enantiomer in the C-2 building block for malyngolide was 98.3% and that of the C-5 building block was 99%. Therefore the enantiomeric excess of the malyngolide is calculated to be (98.3 × 99/100 − 1.7 ×

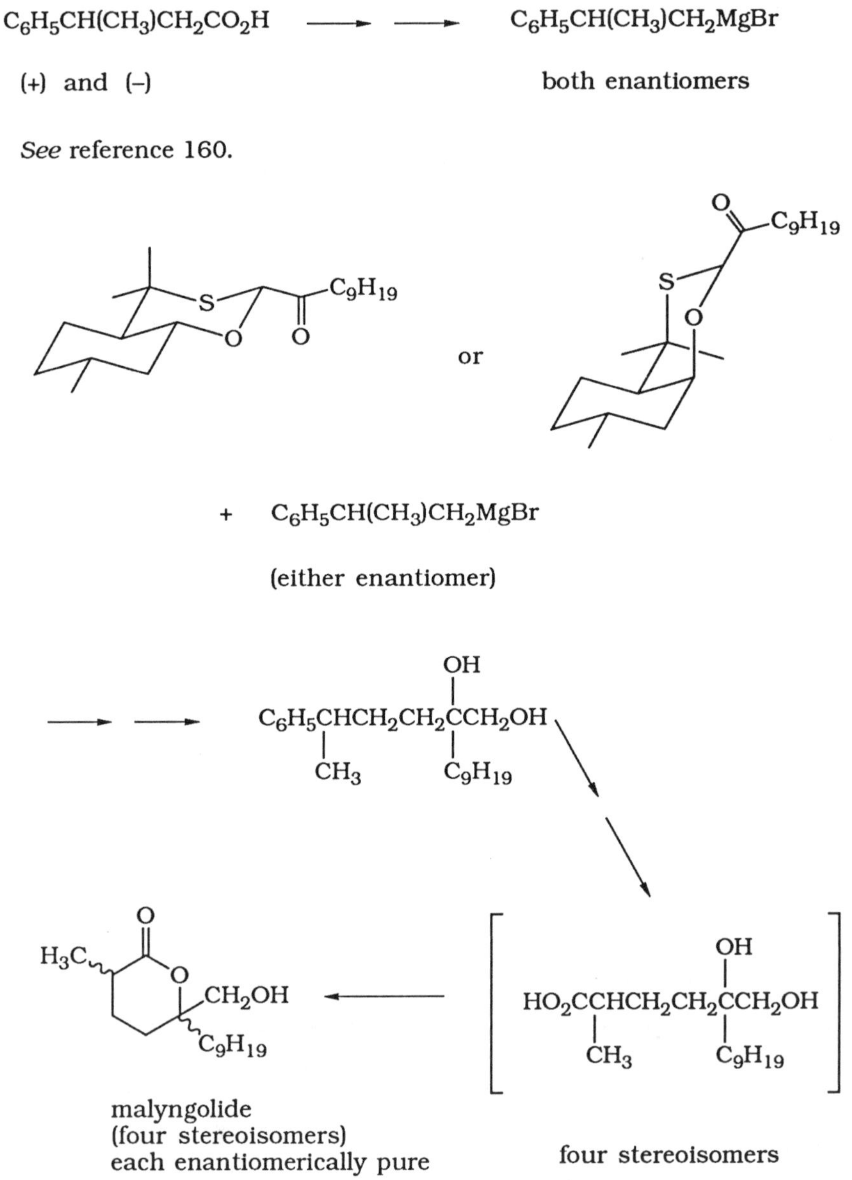

$$C_6H_5CH(CH_3)CH_2CO_2H \longrightarrow \longrightarrow C_6H_5CH(CH_3)CH_2MgBr$$

(+) and (−)   both enantiomers

See reference 160.

or

$$+ \quad C_6H_5CH(CH_3)CH_2MgBr$$

(either enantiomer)

malyngolide
(four stereoisomers)
each enantiomerically pure   four stereoisomers

Scheme 35

1/100)/(98.3 × 99/100 + 1.7 × 1/100) (in which 98.3 × 99/100 is the percentage of malyngolide formed, and 1.7 × 1/100 the percentage of its enantiomer formed) or an astonishing 99.96%! Of course, there is concomitant formation of the diastereomers to the extent of 99 × 1.7/100 + 98.3 × 1/100 = 2.6%. However, the diastereomers can be removed readily by chromatography, and the malyngolide thus obtained is of high diastereomeric, chemical, and enantiomeric purity.

The absence of chemical and diastereomeric impurities can be readily confirmed by proton NMR spectroscopy, but of course, the enantiomeric purity at that high level is virtually impossible to establish experimentally. The question, therefore, is whether the calculated enantiomeric excess is a true figure, that is, whether there is certainty that no racemization had occurred in the synthesis. Now, racemization can occur only if both centers are independently inverted or randomized. Clearly, because such a process affects the two centers independently, it must give rise to epimerization, as well as racemization. Yet it was established experimentally (by chromatography and NMR spectroscopy) that the extent of epimerization was (within experimental limits) equal to that calculated. Therefore no racemization can have occurred, and the enantiomeric excess of the malyngolide is indeed 99.96% (four parts in 10,000 of the wrong enantiomer).

I have mentioned earlier that by the time we had fully developed these synthetic schemes, asymmetric syntheses leading to high enantiomeric excess had become rather commonplace. In particular, Sharpless[140] had developed his excellent synthesis involving enantioselective epoxidation of allylic alcohols, and being catalytic, this method is, in principle, superior to any chemical method involving a chiral auxiliary reagent in stoichiometrical equivalent. It is thus not surprising that a number of the compounds shown in Scheme 34 have also been synthesized asymmetrically by routes involving Sharpless's methodology.

Beyond the syntheses summarized in Scheme 34, the oxathiane work gave rise to two other interesting findings, for neither of which I can claim credit. In several of the syntheses shown in Scheme 34, for example, mevalolactone, citramalic acid, malyngolide, and 6-acetoxy-5-hexadecanolide, the group introduced by the Grignard reaction (cf. Scheme 28, bottom) must be functionalized eventually. In the syntheses just mentioned, the synthon (functional precursor) for the function to be developed (often a carboxyl group) was usually a phenyl group, which was oxidized to $-CO_2H$ at the appropriate stage, usually by means of ruthenium tetroxide. Obviously, it would be preferable to have the functional group present to begin with.

Because alkoxy substituents are compatible with Grignard reactions, Stephen Frye, an excellent graduate student, experimented with side chains in the oxathianyl ketone of the $X-CO-(CH_2)_nOR$ type ($X =$

C-2-substituted oxathiane). The results were fairly dismal; with R = benzyl and n = 1 or 2, diastereomeric excess dropped from the usual 90+% to 33 and 17%, respectively[139,141] (in fact, both additions proceeded contrary to Cram's chelate rule), and even with n = 2 with R = $(C_6H_5)_3C$ or n = 3 or 4 with R = benzyl, stereoselectivity was only 60–75%. Apparently, chelation with the side chain competes with the needed chelation with the oxathiane moiety (Scheme 36).

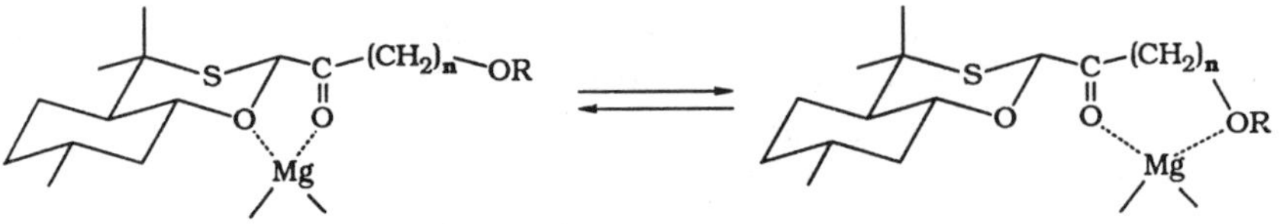

Scheme 36

Mechanistic Studies by Rapid-Injection NMR Spectroscopy and Other Methods

When I presented these facts in a lecture (in Spanish) at the Syntex Laboratories in Mexico City in 1985, Muchowski, the director of research, was in the audience and suggested that we replace the benzyl group by the much bulkier triisopropylsilyl [$(iPr)_3Si$] group. This replacement was done for n = 1 and 2, and the diastereomeric excess indeed came back to 95%.[141] Trimethylsilyl and *tert*-butyldimethylsilyl groups are less effective. In view of the length of the Si–O bond, the effect of $(iPr)_3Si$ may be a relayed electronic effect rather than a direct steric effect.

　　　Stephen Frye was a thoughtful, as well as an ambitious, collaborator. After experiencing the synthetic effectiveness of chelation on several occasions, he asked whether chelation during the course of nucleophilic addition had ever been demonstrated mechanistically. A few papers in the literature show that chelates are formed from α-alkoxy ketones and certain Lewis acids in suitable solvents.[142,143] But no proof exists that such chelates are intermediates in the reaction path of nucleophilic addition rather than shunt complexes. To prove that chelates are intermediates, one must demonstrate that the reaction proceeds faster when chelates are formed than when they are not formed. This proof is difficult to carry out, because the introduction of a chelating alkoxy group in the vicinity of a ketone might have an effect on reaction rate quite apart from its potential for chelation (e.g., an accelerating inductive effect).

Frye recognized that the $(iPr)_3Si$ group might provide a way around this problem, because the inductive effects of $C_6H_5CH_2-O-$ and $(iPr)_3)Si-O$ might be nearly equal. He had just shown that their complexing and chelating abilities were quite different. Therefore, he chose ketones of the type $CH_3CO(CH_2)_nOR$, with $n = 1$ or 2 and R = $C_6H_5CH_2$ or $(iPr)_3Si$, as substrates and $(CH_3)_2Mg$ (to avoid the complicating Schlenk equilibrium) as the reagent. However, these reactions are quite fast. Frye suggested the rapid-injection technique of McGarrity[144] as an appropriate analytical device. In this technique, one of the two solutions to be mixed is contained in the NMR tube in the probe, and the other is injected into the tube while a sequence of single pulses (to record the changing proton spectrum) is initiated to follow the reaction. To the best of our knowledge, however, the only available apparatus was in Lausanne, Switzerland.

Stephen Frye and his wife Susan were willing to travel to Lausanne for 6 months on short notice, and the trip was arranged with the help of Hans Dahn at the University of Lausanne (McGarrity had left), a university off-campus fellowship, and some extra summer support from my grants. The resulting findings[145] were very illuminating. For $n = 1$, rate acceleration (due to chelation) caused the reaction to be too fast to follow even at $-70\,°C$ with the apparatus available at Lausanne, but chelation was demonstrated through a competitive reaction: The substrate with R = $C_6H_5CH_2$ reacted over 100 times faster than that with R = $(iPr)_3Si$. For $n = 2$ the effect was much smaller. The substrate with $n = 2$ and R = $(iPr)_3Si$ reacted at about the same rate as 2-hexanone. However, the substrate with R = $C_6H_5CH_2$ reacted about 2–3 times as fast (acceleration being greatest at the lowest temperature studied), and the reaction was slow enough for its absolute rate constant to be determined. That the effect of chelation via six-membered rings (cf. Scheme 36, $n = 2$ is much less than that via five-membered rings ($n = 1$) is in agreement with much preparative evidence.[110] Finally Frye was able to show that for $n = 1$ the compounds with R = $C_6H_5CH_2$ or $(iPr)_3Si$ react with DIBAL at approximately the same rate as expected on the basis of the known fact (cf. page 95) that DIBAL does not chelate.

The rapid-injection NMR (RINMR) work was carried out at the University of Lausanne, but Dahn was so kind as to give us one of the two injectors extant in his laboratory. With this injector, we were able to continue the work in Chapel Hill. In fact, we have succeeded in studying the absolute rates of the reactions of $CH_3COCH_2OCH_2C_6H_5$ (at -90 and $-70\,°C$) and $CH_3COCH_2OSi[CH(CH_3)_2]$ (at -70 and $-50\,°C$) with $(CH_3)_2Mg$. The second-order rate constant for $CH_3COCH_2OCH_2C_6H_5$ is 0.682/s-M at $-70\,°C$, yet we were able to obtain five kinetic points (making for a good straight-line pseudo-first-order plot) during the first 12 s of reaction. Thus we found the rate ratio (^{k}OCH$_2$C$_6$H$_5$/^{k}OSi(iPr)$_3$) to be 140 at -70 °C and 11 when extrapolated to $+25$ °C.[147] (The improvement is due to the larger size of our computer memory; with these very fast reactions, it is not possible to transfer data to the disk in real time.)

The rapid-injection NMR spectroscopic method promises to serve us in other mechanistic studies and continues to be pursued in our laboratories as this account is being written.

Other Research Topics

The account of the asymmetric synthesis work almost brings this chapter to an end. However, for the record and in fairness to my collaborators, I should mention that enantioselective reactions have by no means been the exclusive topic of interest in my laboratory during the last 10 years. I have already mentioned the work on ^{17}O NMR spectroscopy[101] and some of the recent work in the conformational analysis of carbocyclic[79,148] and heterocyclic[94–99,149,150] compounds.

Carbocyclic Compounds

The work on carbocyclic compounds was done largely with Muthiah Manoharan and Ed Olefirowicz, as well as a Russian group[151] (through collaboration with Nikolai Zefirov of Moscow State University, whose acquaintance I made when he visited the United States in 1979 and spent several weeks in our laboratory). Our studies have involved X-ray structural analysis, as well as molecular-mechanics calculations, for which Ed Olefirowicz's familiarity with the computer proved very helpful. A few of these studies are summarized in Scheme 37.

The first study (Scheme 37, part a)[79b] relates to the conformational equilibrium of phenyl and the geminal phenyl–methyl interaction, a topic previously dealt with by molecular-mechanics calculations by Allinger and Tribble.[152] These authors showed that a major repulsive interaction of axial phenyl is not due to interaction with the *syn*-axial hydrogens but due to the interaction of *ortho* hydrogens of the phenyl group with the adjacent equatorial hydrogens at C-2 and C-6 of the cyclohexane ring. According to their calculations, the phenyl ring turns its flat side toward the *syn*-axial hydrogens; however, a later, more detailed calculation[148] indicates that the phenyl ring actually turns away by 25 ° from this symmetrical position to alleviate the interactions between the *ortho* hydrogens and the equatorial hydrogens.

(a)

When R = H and R' = CH_3, $-\Delta G^\circ = 1.13$ kcal/mol

Hence, $-\Delta G^\circ_{C_6H_5} = 2.87$ kcal/mol

When R = CH_3 and R' = H, $-\Delta G^\circ = -0.32$ kcal/mol

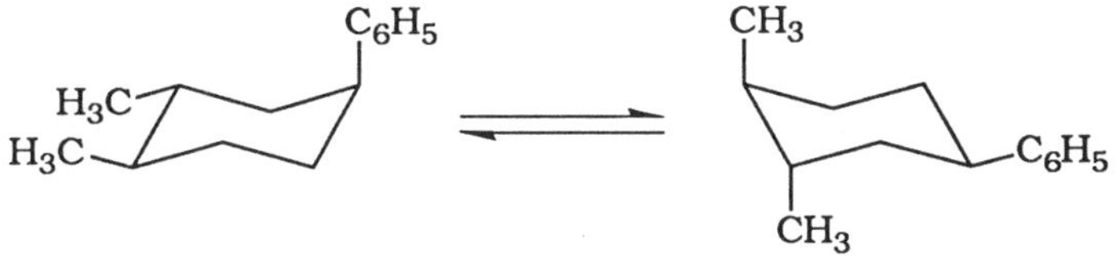

(b)

$-\Delta G = 0.13$ kcal/mol

Hence, CH_3/CH_3 gauche interaction = 0.74 kcal/mol

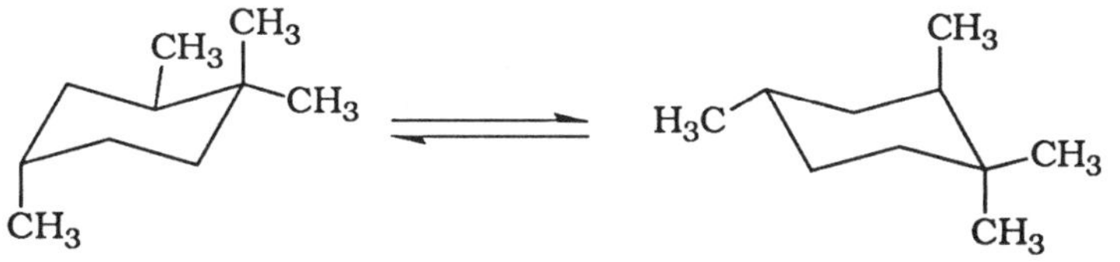

(c)

$-\Delta G = 0.21$ kcal/mol

Scheme 37

The plane of the equatorial phenyl, in contrast, lies in the symmetry plane of the cyclohexane ring. When an axial methyl is now introduced, a severe interaction between two of the methyl hydrogens and one of the phenyl *ortho* hydrogens ensues, and the phenyl turns by ~68 °[148] to avoid this interaction. In fact, the conformation with axial phenyl and equatorial methyl is now the more stable, contrary to what one would expect from the individual conformational energies of phenyl (2.87 kcal/mol)[79b] and methyl (1.74 kcal/mol).[159]

The second study (Scheme 37, part b)[79d] provides a value for the methyl–methyl diequatorial *gauche* interaction (0.74 kcal/mol), and the third study (Scheme 37, part c)[79c] shows that the axial–equatorial equilibrium of a methyl substituent on a cyclohexane ring is only slightly affected by a pair of adjacent geminal methyl substituents, even though in this case, the equatorial methyl has two adjacent *gauche* methyl partners, whereas the axial has only one. This finding confirmed the earlier work of Zefirov's group with polar substituents.[151]

Heterocyclic Compounds

The recent work on heterocyclic compounds[149,150] was done in collaboration with Eusebio Juaristi (a former graduate student with me, now at the Polytechnic Institute in Mexico City) and with a Spanish group at the University (Autónoma) of Madrid under J. L. García, and the University of Seville under Felipe Alcudia, one of my former postdoctoral collaborators. Through the latter collaboration, Ernesto Brunet from Madrid spent some time in my laboratory and discovered a very interesting strong concentration effect on conformation in *cis*-3-hydroxythiane oxide[153] (Scheme 38). At high concentrations in methylene chloride, the diequatorial conformer **38** is strongly favored, presumably because of more favorable intermolecular hydrogen bonding. At low concentrations, the diaxial conformer **37** becomes strongly preferred; in dilute solution, association is not favorable, and only intramolecular hydrogen bonding is possible.

In a solvent that can itself donate and accept hydrogen bonds, such as methanol, **38** is favored at all concentrations. This work complements earlier studies in the phencyclidine series[154] indicating large solvent and temperature effects on the conformational preference of the piperidine moiety (Scheme 39 and Table III).

Other stereochemical studies have involved the configurations of morphine[155] and azabicyclononanes,[156] as well as neighboring-group participation through four-membered rings[157] and stereochemical studies of the 1-phenylcyclohexyl carbanion.[158]

Summary and General Observations

Let me, at this point, summarize my career: what made me into a chemist, how I got started, how one idea and one research problem led to another, and how I achieved success.

I became interested in chemistry through experiments and demonstrations. By the time I read about the subject systematically, I

Ernest L. Eliel at the 1989 Bürgenstock Conference. (Photo taken by K. Zimmermann at the EUCHEM Conference on Stereochemistry, Bürgenstock, 1989, by order of the Organizing Committee.)

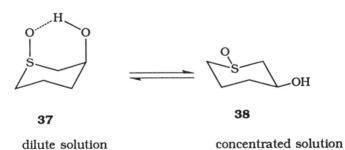

37   **38**

dilute solution   concentrated solution

Scheme 38

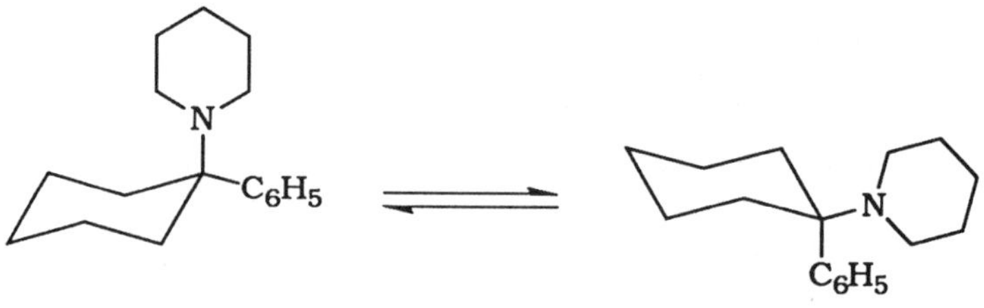

Scheme 39

Table III. Effect of Solvent on Conformational
Preference of Piperidine Moiety

Solvent	$-\Delta G^{\circ}$ (kcal/mol)
CD_3COCD_3–CD_3CN	0.0
CD_2Cl_2	0.45
CD_3OD–CD_2Cl_2 (−80 °C)	1.7
CD_3OD–CD_2Cl_2 (−25 °C)	1.0

was already deeply interested in and committed to it. College in Havana might well have stifled that interest had it not been for the laboratory experience with Rosenkranz and Kaufmann. In graduate school, I was fortunate to work on a very successful, as well as very stimulating, research problem and to be in the presence of a group of able and lively graduate students and young faculty members.

At the beginning of my independent career, I combined some potboiling (continuation of the work on alkylation with amines and quaternary ammonium salts) with a problem that had arisen from my own budding interest in stereochemistry (the question of the optical activity of $C_6H_5CHDCH_3$). This research led to a more general investigation of the stereochemistry of halide reduction with metal hydrides. Although this problem was never solved basically, it led to interesting and, eventually, important research on the properties of "mixed hydrides" (aluminum hydride and aluminum dichlorohydride). Another early problem, the attempted synthesis of yohimbine, although unsuccessful, put me in contact with D. H. R. Barton, indirectly at first via Al Burgstahler (who had initiated the problem as a senior) and directly later. My association with Barton, though at a distance, had an important impact on my later thinking and research.

The hydride work went on for over 20 years; it was nourished by the general interest in metal hydrides during the 1950s and 1960s. Eventually, this field was largely taken over by H. C. Brown, who received the Nobel prize for his discoveries of a facile synthesis of B_2H_6 and its vast potential as a chemical reagent. In the course of the hydride work, I was introduced in 1952 to mass spectrometry. Unfortunately, despite the excitement and the interest this area offered even then, I was unable to work seriously in the field, because my colleagues at Notre Dame were unwilling to let me share their facilities and the expense of the instrumentation put it beyond my own reach.

In 1950, Barton introduced me to conformational analysis, a field that has fascinated me ever since. Our contributions were in the conformational analysis of mobile systems. At first we used the kinetic method based on the Winstein–Holness equation that we developed independently. At the same time (1955), we originated the method of quantitative equilibration, which became a very accurate method of determining configurational and hence conformational equilibria, in part as a result of the development of gas chromatography as an accurate quantitative analytical technique in the late 1950s.

In late 1958, I was introduced to NMR spectroscopy in Jack Roberts's laboratory. This technique has been of immense value in conformational analysis during the last 30 years, its importance increasing with the development of better instruments in the 1960s and especially with ^{13}C NMR spectroscopy becoming readily accessible in the early 1970s. Our principal contributions in the conformational analysis of saturated heterocyclic compounds have depended heavily on low-temperature ^{13}C NMR spectroscopic analysis, although the equilibration method also continues to be useful.

The work with heterocyclic compounds led to the exploration of carbocations of the type $(RO)_2CH^+$ and of carbanions of the type $(RS)_2CH^-$. These studies, in turn, serendipitously led to the discovery of the very stereoselective electrophilic reactions of 2-lithio-1,3-dithianes. On the basis of the discovery that these species react to give nearly exclusively equatorial products, we developed a highly (generally >95%) enantioselective synthesis of the chiral aldehyde synthons RR'COHCHO and RCHOHCHO. This development, in turn, has made possible the synthesis of single enantiomers of a variety of natural products during the last 10 years. However, during this period, many other enantioselective syntheses were developed that compete with ours, notably the now widely used Sharpless epoxidation reaction.

Our research work has been cited widely and thus, evidently, studied widely. I ascribe this popularity of our work, in part, to the clarity of exposition for which I have always striven (and which I learned from my mentor, Harold Snyder) and, in part, to the fact that I have reviewed our work at regular intervals. Once one's reputation is established, it becomes self-reproducing, because once one is known, one's colleagues read one's papers!

My life-long interest in stereochemistry led me to the writing of a book on this subject. This book (published in 1962) became quite important and, incidentally, contributed to my reputation. For some years now, I have struggled to update this book! The book *Conformational Analysis* (1965) and our series *Topics in Stereochemistry* (19 volumes to date, 1967–1989) have also contributed to the development of stereochemistry worldwide.

As is appropriate to this collection of autobiographical accounts, I have traced my career mainly in terms of the progress of my research. Sometimes such progress has been by design, as in the aluminum dichlorohydride mixed-hydride reduction of acetals, the work on quantitative conformational analysis, the early work in proton NMR spectroscopy, and the conformational studies of heterocyclic compounds. In other cases, progress has been made by a stroke of good luck, as in the dithiane carbanion studies, or simply by dead reckoning, as in the mixed-hydride reduction of epoxides. Yet other projects developed from collaborative efforts with others, for example, the ^{13}C NMR studies, and of course in many cases, a combination of two or more of these factors were crucial, as in the work on asymmetric synthesis.

Although I have worked on a number of occasions with other groups—within my own institution, elsewhere in the United States, or abroad (my articles and books have appeared in 12 countries in eight different languages)—most of these efforts have been of short duration. On the whole, I have not enjoyed such collaborative efforts for very long, probably because the main motivation for my research has always been the fascination I saw in the research project itself. If one collaborates, one frequently has to undertake projects other than those one considers most fascinating and most important. For the same reason, I have asked for (and received) relatively little support from the so-called "mission-oriented" agencies, mainly because generally I have not been eager to concern myself with the applications of my research. Most of my research support has come from the National Science Foundation and the Petroleum Research Fund of the American Chemical Society, as well as from other government agencies before, under the impact of the Mansfield amendment in 1970, they became as mission oriented as they are today.

There is, however, one group of collaborators that I have always appreciated in the highest degree and worked with to the very best of my ability, namely the graduate students, postdoctoral associates, and undergraduate researchers in my group. As I have said earlier, the success of a project is intimately and inevitably tied to the ability of the person who executes it in the laboratory. The professor can teach, advise, and inspire, but only the able and motivated student can successfully carry out the work at the bench. I am very grateful to the many able collaborators who have had important roles, over a period of many years, in bringing to fruition the research I have described in these pages.

This brings me to the subject of teaching, about which I have so far said little, although by the rules of the institutions for which I have worked, it occupies approximately half of my time and effort. I have always enjoyed teaching, both because it is important to pass on the wisdom and knowledge one acquires through reading and research and because it hones one's thinking. One cannot teach successfully what one does not oneself understand, and the questions asked by students stimulate one's curiosity. On one or two occasions I have refused to consider academic offers that involved little teaching.

Although I did not start out as a successful teacher (because like many young instructors, I overestimated what could be expected of my students), I improved with time and eventually won two teaching awards, one given nationally by the Manufacturing Chemists' (now Chemical Manufacturers') Association in 1965 and another awarded by the University of North Carolina in 1975. The first award was, I am sure, given for teaching in the widest sense, including not only classroom and laboratory instruction but also the supervision of graduate students and the writing of reviews, educational articles, and textbooks. The second award was mainly for classroom teaching at the undergraduate level, and I must remark sadly that I doubt I could win such an award today, 14 years later. Although I still teach undergraduate courses regularly, I enjoy them less. I find in today's students a lack of ability to reason, to express themselves in writing, and to make good use of supplemental written material. These deficiencies come on top of the perennial problem of the students not having enough time to do justice to their courses.

When I was in Edinburgh, my plan of study for the B.S. Chemistry degree included three courses in the freshman year, two (chemistry and physics or biology) in the sophomore and junior years, and only chemistry in the senior year. History, English, foreign language, geography, etc., had supposedly been learned in high school (as they had been learned in the German *Realgymnasium* mentioned at the beginning of this account). Sadly, the American student lacks this preparation.

Thus, as a member of the university-wide curriculum committee in the late 1970s, I felt impelled to agree to a substantial dose of required courses in English (including literature), foreign language, history, philosophy, fine arts, and social sciences, as well as a minimum of mathematics and science for all students. Unfortunately, in the end, the package had become so large that we had to opt out of part of it for our B.S. students, because it did not leave enough room for the necessary science courses! Clearly, this problem will never be resolved unless we turn an entirely new leaf in our junior high and high school education, at which levels English, history, geography, foreign language, and a good dose of mathematics (including logical thinking) must be taught!

As I implied earlier, I have cherished the many trips abroad I have been able to take, not only to all parts of Europe, but to Japan (where I once gave a 2-week course for which I produced a detailed synopsis of each lecture), China (where my lectures were translated by an interpreter—a disturbing experience at first, because of course, it totally destroys one's style), India, Australia, Brazil, Peru, Argentina, Israel, and several other countries. Such visits, and especially those that extend over more than a few days, give one a wonderful opportunity to observe the customs and, perhaps more important, the thought patterns of other nations and cultures.

I should also say a few words about public service. Perhaps because I came to the United States as an immigrant, I have always felt that I should return something to a society that has given me such great opportunities. Admittedly, that return has been diffuse. Early on, largely under the influence of Charlie Price, I became active in the South Bend chapter of United World Federalists. When I began to feel that this particular involvement was unpromising, I transferred my allegiance to the South Bend International Relations Council, whose

Ernest L. Eliel in Israel in 1967 with the late Gerhard Schmidt (right).

president I became. I am still involved in projects to assist some of the developed and less developed countries of Latin America. Later, under the impetus of my daughter's schooling, I was a member of the board of a private school in South Bend. I have also served as chapter president of Sigma Xi and of the American Association of University Professors, and I have served on committees of the American Association for the Advancement of Science. My fund-raising abilities have been honed by service to the United Jewish Appeal and to the local United Fund, which stands me in good stead because I am presently involved in a major fund drive for the ACS. However, as I mentioned earlier, my major and most consistent and long-standing service has been to my professional society, the ACS. I served as the chairman of the board of directors of the ACS in 1987, 1988, and 1989.

In looking back over my career, I am cognizant of having been very fortunate in that my most productive years fell into a period when organic chemistry was at a peak, when the quality of graduate students and postdoctoral associates in organic chemistry was generally high, and when support for research was ample. Such a period is not likely

In Buenos Aires in 1987 at the headquarters of the Argentine Chemical Association on the occasion of their 75th anniversary. Left to right: Ernest L. Eliel, Juan Rogelio Rodríguez (President of ACA), and Andrés O. M. Stoppani (President of the Argentine Academy of Sciences).

Ernest L. Eliel presenting the 1988 Cope Award to Ken Wiberg at the ACS National Meeting in Los Angeles. (C&E News photo by E. Carpenter.)

to recur, at least not for the core area of organic chemistry. But because I have never believed in the compartmentalization of science, I am not excessively concerned. Chemistry will continue to be the central science, even while its focus is changing. There is much excitement in the application of chemistry to biological problems and also to the properties of materials such as superconductors. Where there is excitement, there will be students and (I hope!) also money.

In conclusion, perhaps I should try to analyze what have been the sources of whatever success I have had. (I do this with considerable hesitation, because I can hear my friend Albert Eschenmoser say, "It is not up to you to say that you have been successful; it is for others to decide if it is so.") Tradition has, no doubt, played an important part. I come from a family in which motivation, combined with honesty and the willingness to work hard, was taken for granted. I took to this tradition instinctively; it never had to be made explicit. The events of my emigration taught me to be resilient and never to succumb to failure. Through conversations with Ludwig Meyer in 1947, I learned that in research persistence in execution is as important as originality in conception. Stephen Kaufmann and Harold Snyder impressed on me the importance of careful experimentation, and Snyder also taught me the importance of clear and organized writing, a lesson that has been invaluable. I am sure that the impact of my writing—be it original articles, reviews, or books—has been enhanced because people found it relatively easy to understand what I had to say.

Snyder also set me on the road of clear oral presentation through his evening research seminars, although I still had some distance to go in this direction when I received my Ph.D. He did, however, impress on me early on (as did Price subsequently and Rossini much later) that the ultimate outcome of research must be publication, and I have always been very cognizant of that necessity, to the point that I would contemplate early in the execution of a research problem how it should be published once it came to fruition. This desire for order in research may have been both salutatory and detrimental. Nature is not always orderly in the way it presents itself, and to try to make it so can lead to oversimplification. But then, when Bob Fuson at the University of Illinois was once asked whether he did not oversimplify the material in his lectures, he responded that any successful teacher must oversimplify. And to the extent that publication is a way of teaching others what we have done, this statement, at least in some degree, applies to research as well.

As I end this story in the year 1990, I have been honored three times: once by my adoptive state of North Carolina, which conferred its

Eva and Ernest in 1989. (Photo by J. I. Seeman.)

1986 Award in Science on me, and again by my colleagues and students who arranged a gala birthday party with a symposium for my friend and colleague Bob Parr and myself on our 65th birthdays. I was most gratified that so many of my former collaborators came back for the birthday celebration; it brought back many happy memories of events chronicled in this account.

And, finally, the University of Notre Dame is conferring an honorary doctorate on me in May 1990. I am particularly pleased by this last honor because I believe that it is given not just for scientific achievement and public service, but also as a recognition that I did contribute to the Notre Dame Chemistry Department in my 24 years there.

References

1. Koch, E. *Deemed Suspect—A Wartime Blunder*; Methuen: New York, 1980.

2. Kaufmann, E.; Eliel, E. L.; Rosenkranz, J. *Ciencia (Mexico)* **1946**, *7*, 136.

3. Eliel, E. L. *J. Chem. Ed.* **1944**, *21*, 583.

4. Zechmeister, L.; Cholnoky, L. *Principles and Practice of Chromatography*; Chapman & Hall: London, 1941.

5. Snyder, H. R.; Smith, C. W. *J. Am. Chem. Soc.* **1944**, *66*, 350.

6. Albertson, N. F.; Archer, S.; Suter, C. M. *J. Am. Chem. Soc.* **1944**, *66*, 500; ibid. **1945**, *67*, 36.

7. Snyder, H. R.; Eliel, E. L. *J. Am. Chem. Soc.* **1948**, *70*, 1703.

8. Snyder, H. R.; Eliel, E. L. *J. Am. Chem. Soc.* **1948**, *70*, 1857.

9. Snyder, H. R.; Eliel, E. L. *J. Am. Chem. Soc.* **1948**, *70*, 3855.

10. Brewster, J. H.; Eliel, E. L. In *Organic Reactions*; Adams, R., Ed.; Wiley: New York, 1953; Vol. 7, p 99.

11. For a much later summary of this work, see Arigoni, D.; Eliel, E. L. In *Topics in Stereochemistry*; Eliel, E. L.; Allinger, N. L., Eds.; Wiley-Interscience: New York, 1969; Vol. 4, p 127.

12. Eliel, E. L.; Peckham, P. E. *J. Am. Chem. Soc.* **1950**, *72*, 1209.

13. Eliel, E. L. *J. Am. Chem. Soc.* **1949**, *71*, 3970.

14. Johnson, J. E.; Blizzard, R. H.; Carhart, H. W. *J. Am. Chem. Soc.* **1948**, *70*, 3664.

15. e.g., Streitwieser, A. *J. Am. Chem. Soc.* 1953, 75, 5014; Loewus, F. A.; Westheimer, F. H.; Vennesland, B. *J. Am. Chem. Soc.* 1953, 75, 5018. See also ref. 11.

16. Eliel, E. L.; Freeman, J. P. *J. Am. Chem. Soc.* 1952, 74, 923.

17. (a) Eliel, E. L.; Herrmann, C.; Traxler, J. T. *J. Am. Chem. Soc.* 1956, 78, 1193; (b) Eliel, E. L.; Prosser, T. J. *J. Am. Chem. Soc.* 1956, 78, 4045; (c) Eliel, E. L.; Traxler, J. T. *J. Am. Chem. Soc.* 1956, 78, 4049; (d) Eliel, E. L.; Delmonte, D. W. *J. Am. Chem. Soc.* 1956, 78, 3226; ibid. 1958, 80, 1744.

18. Eliel, E. L.; McBride, R. T.; Kaufmann, S. *J. Am. Chem. Soc.* 1953, 75, 4291; see also Farley, C. P.; Eliel, E. L. *J. Am. Chem. Soc.* 1956, 78, 3477.

19. Barton, D. H. R. *Experientia* 1950, 6, 316; *Topics in Stereochemistry;* Allinger, N. L.; Eliel, E. L., Eds.; Wiley-Interscience: New York, 1971; Vol. 6, p 1.

20. Prosser, T. J.; Eliel, E. L. *J. Am. Chem. Soc.* 1957, 79, 2544.

21. Eliel, E. L.; Prosser, T. J.; Young, G. W. *J. Chem. Ed.* 1957, 34, 72.

22. Eliel, E. L.; Wilken, P. H.; Fang, F. T. *J. Org. Chem.* 1957, 22, 231.

23. cf. Hassel, O.; Viervoll, H. *Acta Chem. Scand.* 1947, 1, 149; Hassel, O.; Ottar, B. ibid. 1947, 1, 929; See also Hassel, O. In *Topics in Stereochemistry;* Allinger, N. L.; Eliel, E. L., Eds.; Wiley-Interscience: New York, 1971; Vol. 6, p 11.

24. Read, J.; Grubb, W. J. *J. Chem. Soc.* 1934, 1779.

25. (a) Eliel, E. L. *Experientia* 1953, 9, 91. (b) Pehk, T. *Magn. Reson. Relat. Phenom. Proc. Congr. AMPERE 20th* 1978, 496; *Chem. Abstr.* 1980, 93, 186572m.

26. Eliel, E. L. In *Steric Effects in Organic Chemistry;* Newman, M. S., Ed.; Wiley: New York, 1956; p 61.

27. (a) Curtin, D. Y. *Rec. Chem. Prog. Kresge-Hooker Sci. Lib.* 1954, 15, 111. (b) Seeman, J. I. *Chem. Rev.* 1983, 83, 83.

28. Eliel, E. L.; Pillar, C. *J. Am. Chem. Soc.* 1955, 77, 3600.

29. Winstein, S.; Holness, N. J. *J. Am. Chem. Soc.* 1955, 77, 5562.

30. (a) Eliel, E. L.; Ro, R. S. *Chem. Ind. (London)* 1956, 251; (b) idem *J. Am. Chem. Soc.* 1957, 79, 5995.

31. Eliel, E. L.; Lukach, C. A. *J. Am. Chem. Soc.* 1957, 79, 5986.

32. Eliel, E. L.; Allinger, N. L.; Angyal, S. J.; Morrison, G. A. *Conformational Analysis*; Wiley-Interscience: New York, 1965; reprinted by the American Chemical Society: Washington, DC, 1981; footnote on pp 48–49; see also Seeman, J. I. *Chem. Rev.* 1983, *83*, 83.

33. Eliel, E. L.; Ro, R. S. *J. Am. Chem. Soc.* 1957, *79*, 5992.

34. e.g., Meyerson, S.; Rylander, P. N.; Eliel, E. L.; McCollum, J. D. *J. Am. Chem. Soc.* 1959, *81*, 2606; Eliel, E. L.; McCollum, J. D.; Meyerson, S.; Rylander, P. N. *J. Am. Chem. Soc.* 1961, *83*, 2481.

35. cf. Eliel, E. L. *Rec. Chem. Progr. Kresge-Hooker Sci. Lab.* 1961, *22*, 129.

36. Roberts, J. D. *Nuclear Magnetic Resonance*; McGraw-Hill: New York, 1959.

37. e.g., Gutowsky, H. S.; Saika, A. *J. Chem. Phys.* 1953, *21*, 1688.

38. (a) Eliel, E. L. *Chem. Ind. (London)* 1959, 568; (b) *Curr. Contents* 1982, *43*, 22.

39. Eliel, E. L.; Gianni, M. H. *Tetrahedron Lett.* 1962, 97.

40. Eliel, E. L.; Haber, R. G. *J. Am. Chem. Soc.* 1959, *81*, 1249.

41. Eliel, E. L. *J. Chem. Ed.* 1960, *37*, 126.

42. (a) Eliel, E. L.; Schroeter, S. H. J. *J. Am. Chem. Soc.* 1965, *87*, 5031; (b) Eliel, E. L.; Gilbert, E. C. *J. Am. Chem. Soc.* 1969, *91*, 5487.

43. Eliel, E. L.; Brett, T. J. *J. Am. Chem. Soc.* 1965, *87*, 5039.

44. Eliel, E. L.; Doyle, T. W.; Daignault, R. A.; Newman, B. C. *J. Am. Chem. Soc.* 1966, *88*, 1828; Eliel, E. L.; Doyle, T. W. *J. Org. Chem.* 1970, *35*, 2716; Newman, B. C.; Eliel, E. L. *J. Org. Chem.* 1970, *35*, 3641.

45. Eliel, E. L.; Knoeber, M. C., Sr. *J. Am. Chem. Soc.* 1966, *88*, 5347; ibid. 1968, *90*, 3444.

46. Manoharan, M.; Eliel, E. L. *Tetrahedron Lett.* 1984, *25*, 3267.

47. de Kok, A. J.; Romers, C. *Rec. Trav. Chim. Pays-Bas* 1970, *89*, 313.

48. Riddell, F. G.; Robinson, M. J. T. *Tetrahedron* 1967, *23*, 3417.

49. cf. Eliel, E. L. In *Topics in Current Chemistry*; Springer-Verlag: Heidelberg, 1982; Vol. 105, p 1.

50. Mislow, K.; Raban, M. In *Topics in Stereochemistry*; Allinger, N. L.; Eliel, E. L., Eds.; Wiley-Interscience: New York, 1967; Vol. 1, p 1.

51. Eliel, E. L.; Biros, F. J. *J. Am. Chem. Soc.* **1966,** *88,* 3334.

52. Eliel, E. L.; Martin, R. J. L. *J. Am. Chem. Soc.* **1968,** *90,* 682.

53. Eliel, E. L.; Neilson, D. G.; Gilbert, E. C. *Chem. Commun.* **1968,** 360.

54. Eliel, E. L.; Reese, M. C. *J. Am. Chem. Soc.* **1968,** *90,* 1560.

55. Eliel, E. L.; Giza, C. A. *J. Org. Chem.* **1968,** *33,* 3754.

56. cf. Lemieux, R. U. In *Molecular Rearrangements*; de Mayo, P., Ed.; Wiley-Interscience: New York, 1964; Vol. 2, p 709.

57. Altona, C. Ph.D. Dissertation, Leiden, Netherlands, 1964; Romers, C.; Altona, C.; Buys, H. R.; Havinga, E. In *Topics in Stereochemistry*; Eliel, E. L.; Allinger, N. L., Eds.; Wiley-Interscience: New York, 1969; Vol. 4, p 39.

58. Hutchins, R. O.; Kopp, L. D.; Eliel, E. L. *J. Am. Chem. Soc.* **1968,** *90,* 7174.

59. Eliel, E. L.; Kopp, L. D.; Dennis, J. E.; Evans, S. A. *Tetrahedron Lett.* **1971,** 3409.

60. Eliel, E. L. *Kem. Tidskr.* **1969,** No. 6–7, 22.

61. Eliel, E. L. *Acc. Chem. Res.* **1970,** *3,* 1.

62. e.g., Jeffrey, G. A.; Yates, J. H. *J. Am. Chem. Soc.* **1979,** *101,* 820.

63. Eliel, E. L.; Hutchins, R. O. *J. Am. Chem. Soc.* **1969,** *91,* 2703.

64. Eliel, E. L.; Kaloustian, M. K. *Chem. Commun.* **1970,** 290.

65. Willy, W. E.; Binsch, G.; Eliel, E. L. *J. Am. Chem. Soc.* **1970,** *92,* 5394.

66. Eliel, E. L. *Pure Appl. Chem.* **1971,** *25,* 509.

67. Eliel, E. L.; Nader, F. W. *J. Am. Chem. Soc.* **1969,** *91,* 356; ibid. **1970,** *92,* 584.

68. Fraser, R. R.; Schuber, F. J. *Can. J. Chem.* **1970,** *48,* 633; Durst, T.; Fraser, R. R.; McClory, M. R.; Swingle, R. B.; Viau, R.; Wigfield, Y. Y. *Can. J. Chem.* **1970,** *48,* 2148.

69. Hartmann, A. A.; Eliel, E. L. *J. Am. Chem. Soc.* **1971,** *93,* 2572; Eliel, E. L.; Hartmann, A. A.; Abatjoglou, A. G. *J. Am. Chem. Soc.* **1974,** *96,* 1807.

70. Abatjoglou, A. G.; Eliel, E. L.; Kuyper, L. F. *J. Am. Chem. Soc.* **1977,** *99,* 8262.

71. Lehn, J.-M., personal communication.

72. Lehn, J.-M.; Wipff, G. *J. Am. Chem. Soc.* **1976**, *98*, 7498.

73. Amstutz, R.; Dunitz, J. D.; Seebach, D. *Angew. Chem. Int. Ed. Engl.* **1981**, *20*, 465.

74. cf. Eliel, E. L. *Tetrahedron* **1974**, *30*, 1503.

75. Dalling, D. K.; Grant, D. M. *J. Am. Chem. Soc.* **1967**, *89*, 6612.

76. Jones, A. J.; Eliel, E. L.; Grant, D. M.; Knoeber, M. C.; Bailey, W. F. *J. Am. Chem. Soc.* **1971**, *93*, 4772.

77. Eliel, E. L.; Bailey, W. F.; Kopp, L. D.; Willer, R. L.; Grant, D. M.; Bertrand, R.; Christensen, K. A.; Dalling, D. K.; Duch, M. W.; Wenkert, E.; Schell, F. M.; Cochran, D. W. *J. Am. Chem. Soc.* **1975**, *97*, 322.

78. Berlin, A. J.; Jensen, F. R. *Chem. Ind. (London)* **1960**, 998.

79. (a) Eliel, E. L.; Kandasamy, D. *J. Org. Chem.* **1976**, *41*, 3899; (b) Eliel, E. L.; Manoharan, M. *J. Org. Chem.* **1981**, *46*, 1959; (c) Eliel, E. L.; Chandrasekaran, S. *J. Org. Chem.* **1982**, *47*, 4783; (d) Manoharan, M.; Eliel, E. L. *Tetrahedron Lett.* **1983**, *24*, 453.

80. cf. Eliel, E. L.; Vierhapper, F. W. *J. Am. Chem. Soc.* **1975**, *97*, 2424.

81. cf. Anet, F. A. L.; Yavari, I. *J. Am. Chem. Soc.* **1977**, *99*, 2794.

82. Blackburne, I. D.; Katritzky, A. R.; Takeuchi, Y. *Acc. Chem. Res.* **1975**, *8*, 300.

83. Lambert, J. B.; Featherman, S. I. *Chem. Rev.* **1975**, *75*, 611.

84. Eliel, E. L.; Hofer, O. *J. Am. Chem. Soc.* **1973**, *95*, 8041.

85. Hofer, O.; Eliel, E. L. *J. Am. Chem. Soc.* **1973**, *95*, 8045.

86. Vierhapper, F. W.; Eliel, E. L. *J. Am. Chem. Soc.* **1974**, *96*, 2256; idem *J. Org. Chem.* **1975**, *40*, 2729.

87. Vierhapper, F. W.; Eliel, E. L. *J. Org. Chem.* **1975**, *40*, 2734.

88. Eliel, E. L.; Vierhapper, F. W. *J. Am. Chem. Soc.* **1974**, *96*, 2257.

89. Crowley, P. J.; Robinson, M. J. T.; Ward, M. G. *J. Chem. Soc., Chem. Commun.* **1974**, 825; idem *Tetrahedron* **1977**, *33*, 915.

90. Eliel, E. L.; Vierhapper, F. W. *J. Org. Chem.* **1976**, *41*, 199.

91. Vierhapper, F. W.; Eliel, E. L. *J. Org. Chem.* **1977**, *42*, 51.

92. Vierhapper, F. W.; Eliel, E. L. *J. Org. Chem.* **1979**, *44*, 1081.

93. Hargrave, K. D.; Eliel, E. L. *Tetrahedron Lett.* **1979**, 1987; idem *Isr. J. Chem.* **1980**, *20*, 127.

94. Willer, R. L.; Eliel, E. L. *J. Am. Chem. Soc.* **1977**, *99*, 1925.

95. Eliel, E. L.; Kandasamy, D. *Tetrahedron Lett.* **1976**, 3765.

96. Eliel, E. L.; Kandasamy, D.; Yen, C.-y.; Hargrave, K. D. *J. Am. Chem. Soc.* **1980**, *102*, 3698.

97. (a) Eliel, E. L.; Hargrave, K. D.; Pietrusiewicz, K. M.; Manoharan, M. *J. Am. Chem. Soc.* **1982**, *104*, 3635; (b) Eliel, E. L.; Pietrusiewicz, K. M. *Pol. J. Chem.* **1981**, *55*, 1265.

98. Eliel, E. L.; Willer, R. L. *J. Am. Chem. Soc.* **1977**, *99*, 1936.

99. Eliel, E. L.; Yen, C.-y.; Zúñiga Juaristi, G. *Tetrahedron Lett.* **1977**, 2931.

100. Eliel, E. L.; Rao, V. S.; Riddell, F. G. *J. Am. Chem. Soc.* **1976**, *98*, 3583; Willer, R. L.; Eliel, E. L. *Org. Magn. Reson.* **1977**, *9*, 285; Eliel, E. L.; Rao, V. S.; Pietrusiewicz, K. M. ibid. **1979**, *12*, 461; Eliel, E. L.; Manoharan, M.; Pietrusiewicz, K. M.; Hargrave, K. D. ibid. **1983**, *21*, 94.

101. Eliel, E. L.; Pietrusiewicz, K. M.; Jewell, L. M. *Tetrahedron Lett.* **1979**, 3649; Eliel, E. L.; Liu, K.-T.; Chandrasekaran, S. *Org. Magn. Reson.* **1983**, *21*, 179; Manoharan, M.; Eliel, E. L. *Org. Magn. Reson.* **1985**, *23*, 225; Eliel, E. L.; Chandrasekaran, S.; Carpenter, L. E.; Verkade, J. G. *J. Am. Chem. Soc.* **1986**, *108*, 6651.

102. Eliel, E. L.; Pietrusiewicz, K. M. In *Topics in Carbon-13 NMR Spectroscopy*; Levy, G. C., Ed.; Wiley: New York, 1979; Vol. 3, p 171.

103. cf. Eliel, E. L. *CHEMTECH* **1974**, 758.

104. cf. *van't Hoff-Le Bel Centennial*; Ramsay, O. B., Ed.; ACS Symposium Series 12; American Chemical Society: Washington, DC, 1975.

105. Hermans, P. H. *Z. Physik. Chem.* **1924**, *113*, 337.

106. Eliel, E. L.; Koskimies, J. K.; Lohri, B. *J. Am. Chem. Soc.* **1978**, *100*, 1614.

107. Brown, H. C.; Zweifel, G. *J. Am. Chem. Soc.* **1961**, *83*, 486.

108. a) e.g., Meyers, A. I.; Knaus, G.; Kamata, K. *J. Am. Chem. Soc.* **1974**, *96*, 268. b) e.g., Knowles, W. S.; Sabacky, M. J.; Vineyard, B. D.; Weinkauff, D. J. ibid. **1975**, *97*, 2567.

109. See, for example, the five-volume series, *Asymmetric Synthesis*; Morrison, J. D., Ed.; Academic: New York, 1983–1985.

110. Morrison, J. D.; Mosher, H. S. *Asymmetric Organic Reactions*; Prentice-Hall: Englewood Cliffs, NJ, 1971; reprinted by the American Chemical Society: Washington, DC, 1976.

111. Koskimies, J. Ph.D. Dissertation, University of North Carolina, Chapel Hill, 1976.

112. Cram, D. J.; Kopecky, K. R. *J. Am. Chem. Soc.* **1959**, *81*, 2748; cf. Eliel, E. L. In *Asymmetric Synthesis*; Morrison, J. D., Ed.; Academic: New York, 1983; Vol. 2, p 125.

113. Omura, K.; Sharma, A. K.; Swern, D. *J. Org. Chem.* **1976**, *41*, 957.

114. Eliel, E. L.; Morris-Natschke, S. *J. Am. Chem. Soc.* **1984**, *106*, 2937.

115. Lynch, J. E.; Eliel, E. L. *J. Am. Chem. Soc.* **1984**, *106*, 2943.

116. Corey, E. J.; Erickson, B. W. *J. Org. Chem.* **1971**, *36*, 3553.

117. (a) Inch, T. D.; Ley, R. V.; Rich, P. *J. Chem. Soc. C* **1968**, 1693; (b) Kraus, G. A.; Roth, B. *J. Org. Chem.* **1980**, *45*, 4825.

118. Eliel, E. L.; Frazee, W. J. *J. Org. Chem.* **1979**, *44*, 3598.

119. Eliel, E. L.; Lynch, J. E.; Kume, F.; Frye, S. V. *Org. Synth.* **1987**, *65*, 215.

120. Ko, K.-Y.; Frazee, W. J.; Eliel, E. L. *Tetrahedron* **1984**, *40*, 1333.

121. Eliel, E. L.; Lynch, J. E. *Tetrahedron Lett.* **1981**, *22*, 2855.

122. Eliel, E. L.; Soai, K. *Tetrahedron Lett.* **1981**, *22*, 2859.

123. Eliel, E. L.; Koskimies, J. K.; Lohri, B.; Frazee, W. J.; Morris-Natschke, S.; Lynch, J. E.; Soai, K. In *Asymmetric Reactions and Processes in Chemistry*; Eliel, E. L.; Otsuka, S., Eds.; ACS Symposium Series 185; American Chemical Society: Washington, DC, 1982; pp 37–53.

124. Evans, R. J. D.; Landor, S. R. *J. Chem. Soc.* **1965**, 2553; Evans, R. J. D.; Landor, S. R.; Taylor Smith, R. *J. Chem. Soc.* **1963**, 1506; see also Eliel, E. L. *Tetrahedron Lett.* **1960**, *8*, 16.

125. Prelog, V. *Helv. Chim. Acta* **1953**, *36*, 308.

126. Baldwin, J. W., personal communication, 1982; cf. Schweiter, M. J.; Sharpless, K. B. *Tetrahedron Lett.* **1985**, *26*, 2543.

127. Runge, W. In *The Chemistry of the Allenes*; Landor, S. R., Ed.; Academic: New York, 1982; Vol. 3, p 579.

128. e.g., Caporusso, A. M.; Zoppi, A.; Da Settimo, F.; Lardicci, L. *Gazz. Chim. Ital.* **1985**, *115*, 293.

129. Caporusso, A. M.; Rosini, C.; Lardicci, L.; Polizzi, C.; Salvadori, P. *Gazz. Chim. Ital.* **1986**, *116*, 467; Elsevier, C. J.; Mooiweer, H. H. *J. Org. Chem.* **1987**, *52*, 1536.

130. Eliel, E. L.; Lynch, J. E. *Tetrahedron Lett.* **1987**, *28*, 4813.

131. Frye, S. V.; Eliel, E. L. *J. Org. Chem.* **1985**, *50*, 3402.

132. Frye, S. V.; Eliel, E. L. *Tetrahedron Lett.* **1985**, *26*, 3907; idem *Croat. Chem. Acta* **1985**, *58*, 647.

133. Ohwa, M.; Kogure, T.; Eliel, E. L. *J. Org. Chem.* **1986**, *51*, 2599.

134. Ohwa, M.; Eliel, E. L. *Chem. Lett.* **1987**, 41.

135. Ko, K.-Y. Ph.D. Dissertation, University of North Carolina, Chapel Hill, 1985.

136. Kogure, T.; Eliel, E. L. *J. Org. Chem.* **1984**, *49*, 576.

137. Eliel, E. L.; Alvarez, M. T.; Lynch, J. E. *Nouv. J. Chim.* **1986**, *10*, 749.

138. Ko, K.-Y.; Eliel, E. L. *J. Org. Chem.* **1986**, *51*, 5353.

139. Eliel, E. L. *Phosphorus Sulfur* **1985**, *24*, 73.

140. Katsuki, T.; Sharpless, K. B. *J. Am. Chem. Soc.* **1980**, *102*, 5974.

141. Frye, S. V.; Eliel, E. L. *Tetrahedron Lett.* **1986**, *27*, 3223; idem *J. Am. Chem. Soc.* **1988**, *110*, 484.

142. e.g., Keck, G. E.; Castellino, S. *J. Am. Chem. Soc.* **1986**, *108*, 3847; idem *Tetrahedron Lett.* **1987**, *28*, 281.

143. e.g., Reetz, M. T.; Kesseler, K.; Schmidtberger, S.; Wenderoth, B.; Steinbach, R. *Angew. Chem. Suppl.* **1983**, 1511; Reetz, M. T.; Hüllmann, M.; Seitz, T. *Angew. Chem. Int. Ed. Engl.* **1987**, *26*, 477.

144. McGarrity, J. F.; Prodolliet, J.; Smyth, T. *Org. Magn. Reson.* **1981**, *17*, 59.

145. Frye, S. V.; Eliel, E. L.; Cloux, R. *J. Am. Chem. Soc.* **1987**, *109*, 1862.

146. e.g., Mead, K.; MacDonald, T. L. *J. Org. Chem.* **1985**, *50*, 422.

147. Frye, S. V., unpublished observations.

148. Hodgson, D. J.; Rychlewska, U.; Eliel, E. L.; Manoharan, M.; Knox, D. E.; Olefirowicz, E. M. *J. Org. Chem.* **1985**, *50*, 4838.

149. García-Ruano, J. L.; Rodríguez, J.; Alcudia, F.; Llera, J. M.; Olefirowicz, E. M.; Eliel, E. L. *J. Org. Chem.* **1987**, *52*, 4099.

150. Juaristi, E.; Martínez, R.; Méndez, R.; Toscano, R. A.; Soriano-García, M.; Eliel, E. L.; Petsom, A.; Glass, R. S. *J. Org. Chem.* **1987**, *52*, 3806.

151. Mursakulov, I. G.; Ramazanov, E. A.; Samoshin, V. V.; Zefirov, N. S.; Eliel, E. L. *J. Org. Chem. USSR (Engl. Transl.)* **1979**, *15*, 2186; Mursakulov, I. G.; Ramazanov, E. A.; Guseinov, M. M.; Zefirov, N. S.; Samoshin, V. V.; Eliel, E. L. *Tetrahedron* **1980**, *36*, 1885.

152. Allinger, N. L.; Tribble, M. T. *Tetrahedron Lett.* **1971**, 3259.

153. Eliel, E. L.; Brunet, E. *Tetrahedron Lett.* **1985**, *26*, 3421; Brunet, E.; Eliel, E. L. *J. Org. Chem.* **1986**, *51*, 677.

154. Manoharan, M.; Eliel, E. L.; Carroll, F. I. *Tetrahedron Lett.* **1983**, *24*, 1855. See also Liu, K.-T.; Eliel, E. L. *Heterocycles* **1982**, *18*, 51.

155. Eliel, E. L.; Morris-Natschke, S.; Kolb, V. M. *Org. Magn. Reson.* **1984**, *22*, 258.

156. Jeyaraman, R.; Jawaharsingh, C. B.; Avila, S.; Ganapathy, K.; Eliel, E. L.; Manoharan, M.; Morris-Natschke, S. *Heterocycl. Chem.* **1982**, *19*, 449; Eliel, E. L.; Manoharan, M.; Hodgson, D. J.; Eggleston, D. S. *J. Org. Chem.* **1982**, *47*, 4353.

157. Eliel, E. L.; Knox, D. E. *J. Am. Chem. Soc.* **1985**, *107*, 2946; Eliel, E. L.; Clawson, L.; Knox, D. E. *J. Org. Chem.* **1985**, *50*, 2707.

158. Keys, B. A.; Eliel, E. L.; Juaristi, E. *Isr. J. Chem.* **1989**, *29*, 171.

159. Booth, H.; Everett, J. R. *J. Chem. Soc. Chem. Commun.* **1976**, 278.

160. Mukaiyama, I.; Iwasawa, N. *Chem. Lett.* **1981**, 913.

Index

Index

Production: Paula M. Bérard
Indexing: Colleen P. Stamm
Acquisition: Robin Giroux

Printed and bound by Maple Press, York, PA

Paper meets minimum requirements of American National Standard for Information Sciences—Permanence of Paper for Printed Library Materials, ANSI Z39.48–1984   ∞